A Canticle of Reality

Ishi Nobu

A Canticle of Reality

by Ishi Nobu

A *Spokes of the Wheel* Book

ISBN 978-1-948627-25-2

Web site: *www.ishinobu.com*
Email: *comment@ishinobu.com*

Produced in the United States of America.
1st edition.

Other books by Ishi Nobu:

Besting Yourself: Gaining Power
The Story of Humanity: Ecology & Consequence
The Red Pill: Mastering The Matrix
Clarity: The Path Inside
Unraveling Reality: Beyond the Veil of Existence
Spokes of the Wheel:
- *The Science of Existence*
- *The Web of Life*
- *The Elements of Evolution*
- *The Ecology of Humans*
- *The Echoes of the Mind*
- *The Fruits of Civilization*
- *The Pathos of Politics*
- *The Hub of Being*

Table of Contents

Preface

Spirituality awakens with wonder: when awareness questions its habitat. Curiosity expresses spirituality. Spirituality – desire for awareness – is an inborn trait.

Spirituality dies when wondering ceases. Those who think they know what is going on are the walking dead.

To entrance commoners, ancient spiritual texts were often poetic prose. Parable and analogy were common narrative devices. This followed oral storytelling tradition. The currents of lessons underlying the folksy fables ran deep.

Intellectually, spirituality evolved into philosophy: defining terms and linking concepts into a collage of comprehension.

Advancing technology pushed philosophy into the background. Science came to the fore.

But science proved a supplement to philosophy, not a substitute. A theory is a philosophic construct with facts treated as axioms.

Selfsame questions provoke science, philosophy, and spirituality. What is reality? What is the nature of Nature? How to reconcile the mind with matter? Why am I here? How do I best conduct my life? Is there life after death?

Science has acquired a bevy of amazing discoveries. These revelations provide puzzle pieces to answer the elemental questions and thereby assist in creating a consilience of guidance for living and endeavor: a proper way to be and do.

This is a philosophy book, written in the style of archaic spiritual texts. The ample glossary testifies to terms defined. Underlying the poetic verses is science, with conceptual currents running deep.

Paradox

逆説

Paradox is the nature of Nature.
From chaos comes order.
Longing is for contentment.
Dying begets birth.
Vitality depends upon stillness.
Perception is deception.
The physical is metaphysical.
The absolute becomes relative.
Solid is fluid.
From nothing comes everything.
Diversity disguises unicity.
Reality cannot be.

Commentary

Paradox is the nature of Nature.

A *paradox* is a contradiction resolved by construing a complement. A *complement* is something which completes or balances a system. A paradox conceptually sustains an abiding polar tension.

陰陽

Inyō is Japanese for complementary opposites which comprise an interconnected dynamic. A better-known word in English is *yinyang*. Inyō and yinyang have the same ideograms.

An *ouroboros* is an ancient symbol of a dragon or serpent eating its own tail. An ouroboros traditionally signifies the eternal cycle of incarnation. As a gnostic and alchemical symbol, ouroboros expressed the unity of all things, material and spiritual, which perpetually change through an eternal cycle of creation and destruction.

To recap: a *paradox* is a conceptual system that holds polar constructs in enduring tension. An *inyō* is a dynamic with entangled opposites. An *ouroboros* is an endless cycle of instantiations.

Nature is the perceived exhibition of existence. To understand that exhibit, the mind construes contrasts via comparison. Via comparative contrasts categorization arises. Categorization affords recognition. As such, paradox, inyō, and ouroboros are integral facets of comprehension.

From chaos comes order.

Chaos and order are perceptual inyō. Perception crafts comprehension: molding chaotic sensation into meaningful information. In relying upon conceptual poles for recognition, perception works via paradox.

Longing is for contentment.

Longing is a synonym for desire. *Contentment* is sated desire. The desire cycle is an ouroboros.

As an inyō, desire defines living. Only by and through desire do we live and want to continue to do so. Indeed, the primal desire is for the experience of living.

Inyō underlies desire in what wants are pursued and those which are forgone.

Dying begets birth.

Living is an endless ouroboros. Consciousness is an incarnation of an eternal soul which cycles through lifetimes. A new life begins after the old one ceases to be.

In *transmigration*, the inyō tension dynamics of a previous life are reflectively carried on in the next. This provides for evolutionary continuity.

Transmigration has been a spiritual tenet of most cultures worldwide since antiquity. The term *reincarnation* is used for someone who remembers a previous incarnation. Abundant evidence of transmigration comes from reports of reincarnated children who recall places and speak languages that they could not possibly otherwise know.

Vitality depends upon stillness.

Consciousness is the faculty for awareness. Though consciousness is an interested witness, awareness itself is quiet, like still water. Mentation generates waves upon that stillness. As witnessing requires a show, consciousness and perception are a paradox of tranquility and activity.

The vital energy for living emanates from the will of consciousness, which instantiates in the mental process termed *willmind.* The mind powers itself by the energy of consciousness. The willmind process is an inyō.

Perception is deception.

You experience Nature as a *dualism.* Your mind informs you that you are an individual amid an external world. As reality is instead a monism, dualism is deceit.

Dualism is proven untrue because there is no way to reconcile intangible consciousness and mentation with tangible physicality. A mix of mind and matter is inexplicable. No one in history has devised a science that upholds dualism. Despite this, many researchers, lacking a grounding in philosophy, ignore this seminal issue and blithely assume dualism.

The duplicity of dualism has been known to sages since antiquity. Sagacity is a product of high awareness, which may be starkly contrasted to the iğnorance that defines the lowest level of consciousness. Possessing inner clarity, sages perceive past the deceit of the mind.

The mind's dualism deceit presents a paradox of why it arises. That paradox is resolved by understanding that dualism imparts a sense of realism that is essential to treating living as a meaningful experience. Why deception is integral to living is elaborated in the "Perception" chapter.

With dualism downed, existence must emerge from a singular reality: a *monism*. Either matter makes the mind (matterism) or the mind makes matter (energyism).

Science has thoroughly disproven matterism: the idea of matter constructing consciousness and mentation. Like the mind-body problem that dooms dualism, no one can explain how matter could concoct the intangibility of awareness or the intricacy of mental activity within us which is verifiably certain.

With matterism dispatched as a ruse, the opposite must be true. The mind manufactures the subterfuge of tangible materiality: energyism.

The physical is metaphysical.

The central tenet of *energyism* is that the material world is a mirage of the mind. Physics avers energyism in 2 distinct ways.

The creation of energy from matter and dissolution of matter into energy are processes respectively termed *fusion* and *fission*.

Stars turn matter into energy by fusing atoms, creating tremendous energy which lights and warms the planets which orbit them via radiation. Stars prodigiously feed off hydrogen, the simplest element. Atomically, hydrogen is the seed of all matter. Stars manufacture heavier elements through fusion.

Conversely, matter dies through fission. Atoms release tremendous energy when they decay, as atomic bombs explosively testify.

Matter is born of energy and releases energy thunderously in its death throes. By the paradoxical processes of fusion and fission we see that matter and energy are an ouroboros.

Energy is an inyō in being measurable only by its perceptible effect on matter. That matter is made of energy and energy is only known via matter is a scientific circularity. Squaring that circle into physicality as reality is as impossible as the ancient Greek geometry problem from which the idiom arose.

Working with the geometry of polygons, some ancient Greek mathematicians convinced themselves that a circle could be squared. In the late 5th century BCE, Antiphon argued that inscribing regular polygons within a circle and repeatedly doubling the number of sides eventually fills up the area of the circle. This mathematical technique is known as the *method of exhaustion*.

Any polygon can be squared: a polygon reshaped so that its area matches that of a square. Exhaustion suggests a circle can be squared.

Even the renowned English physicist Isaac Newton believed so. Newton discovered calculus, part of which is geometric integration (integral calculus).

The impossibility of squaring the circle was proven in 1882 with the observation that pi (π) is a transcendental number. π is the ratio of a circle's circumference to its diameter. Since π is not the root of any polynomial with rational coefficients, no dimension can be given to a square that would have an identical area of a circle.

Every measure is mathematical. And what is mathematics but a system of symbols?! Looking behind the numbers related to physicality resolutely resolves to ideas. The same may be said of whatever comes to mind.

In explaining his now-proven equation of matter-energy equivalence ($E=mc^2$), physicist Albert Einstein declared "Everything is energy and that's all there is to it." Physicists readily admit that energy itself is nothing but an idea. Carlos Calle plainly stated that "Energy is an abstract concept."

There is no escaping that energy is entirely conceptual, and thereby, matter is a perceptual ruse of the first order. The primal paradox of existence is that the physical is metaphysical.

The absolute becomes relative.

This inyō indicates that an absolute reality creates the relativity that characterizes existence. Whereas reality is absolute, the perception of Nature is of relatives, by way of contrasting conceptual absolutes. Ideas themselves are purities which get tainted by association with their opposites.

Modern physics emerged by creating an inyō of classical physics. Isaac Newton, classical master, famously declared space and time "absolute," albeit conceding that they are perceived as relative.

Einstein diminished space and time into perceptual relatives. Einstein's relativity theories have been experimentally verified at the cosmic scale. What relativity means is that perception is integral to Nature itself.

Though germane, physics is an aside from the intent of this passage, which intends that the transaction by which existence is fabricated moment by moment is by an "absolute" becoming "relative." Whereas reality is absolute, existence – which is gleaned by perceiving Nature – is relative.

Solid is fluid.

The mind portrays a world of particulates. This ruse is belied by the ongoing construction of existence, which transpires via moment-by-moment assembly from the quantum level up in a symphonic coherence. Continuous process belies the appearance of solidity.

Perception is deception in the large by obscuring the monism of reality for an existential dualism. In the small, perception deceives in depicting a world of objects when, instead, Nature is a flow of activity.

As all is process, the term "mind" is a misnomer. There is no mind. There is instead mentation: the effervescent spinning of symbolic fields.

The common tongue reflects a culture of iğnorance, which is obsessed with objects. When "mind" is used herein, translate the word to "mental behavior."

From nothing comes everything.

This passage posits an ouroboric monism: that the phenomena of existence ("everything") emerge from a noumenon ("nothing"), which is a singular nonexistent force. This monism gives rise to the mental presentation of dualism.

Diversity disguises unicity.

Underlying the diversity which characterizes existence is a reality of oneness: a monism. This is the fundamental inyō which is also a paradox and an ouroboros.

Reality cannot be.

Reality is what is needed for existence to manifest. But reality is distinct from existence. Reality is outside of existence. Reality does not exist.

The double entendre "be" in "reality cannot *be*" refers to 2 expressions of existence: *beings*, which are living entities, and *beĩng*, which is pure awareness. Beings have beĩng.

There are no beings in reality. God is an impossibility. A popular fantasy, the idea of an incorporeal supreme being is absurd. No one has ever had a decent scientific explanation for a God mechanism (how there could be a God), nor why there would be a God, nor why there would be just one God instead of many: monotheism versus polytheism.

The notion that God judges human morality but does nothing about its transgression is ridiculous. The twist that consenting sexual acts have moral implication is perverse: a philosophic obscenity. Knowing the evolution of the myth of God reveals the ruse at its roots.

Monotheism was enthusiastically approved by authorities as a means of social control. Moralizing ideologies gelled across the world during the Axial Age, between the 8th and 3rd centuries BCE, during an early wave of global conquests.

Moralizing religion came as states evolved from the tribal level to culturally diverse societies. The idea of a judgmental God was sold by the dominant tribe to pacify conquered subordinates of different cultures. Violent victory resulted in dogmatic false morality that had as its prime objective obedience to authority. That it worked so well testifies to the gullibility and domestication of ordinary people (the Collective).

The reality/existence axis is paradoxical, as are the inyō dynamics that underlie physics, chemistry, and biology. The next chapter further distinguishes the conceptual poles of reality and existence.

Existence

存在

Reality is that which necessarily is.

Reality orders existence. Sans existence, nothing is.

Existence requires contrast. Sans contrast, there can be no diversity. Sans diversity, there is no existence.

Existence requires consciousness. Sans awareness, nothing can exist.

Existence requires life. Sans life, there are no vessels for consciousness.

Existence requires coherence. Sans order, there can be no comprehension. Sans comprehension, awareness is useless.

Existence requires Nature. Sans perception of expression, nothing exists.

Existence requires credibility. Sans credibility, Life loses purpose. Sans purpose, Life loses meaning. Sans meaning, there is no jeopardy. Jeopardy grants credibility.

Existence is all that is.

Commentary

> Reality is that which necessarily is.

Reality is what must be for existence to transpire. Reality is the ordering principle for existence.

A *noumenon* signifies nothingness: outside of existence. Reality is noumenal. What exists are *phenomena*: whatever may be perceived and thereby experienced.

Reality and existence are processes. Their mechanics are parametric process mechanisms. What we perceive as objects are really processes with perceptual form stability.

Cöcö is reality. *Cöcö* is an acronym for the unitary field of Cönsciousness and cöherence. Cöcö mechanics are described in the next chapter.

Reality orders existence.

The word "orders" here has double meaning. Cöcö both demands ("orders") existence and sets the operational arrangement of ("orders") existence.

The rest of this chapter defines existence through its requirements. The mind construes *existence* as actuality.

Sans existence, nothing is.

Existence is what *is*. Without existence, there is no *is*. Reality is necessary for existence to manifest. Lacking existence, nothing is necessary. Without existence, there would be no reality. Reality and existence are a codependent paradox: an inyō – and, as an eternal cycling process, an ouroboros.

Existence requires contrast. Sans contrast, there can be no diversity. Sans diversity, there is no existence.

This passage states that perceptual differentiation – "contrast" – creates the diversity that is necessary for the perceptual uptake we call *existence*. Existence requires the ability to perceive diversity, which is a process done via contrast.

Existence requires consciousness. Sans awareness, nothing can exist.

Consciousness is the ability to be aware. Without awareness, nothing would exist because there would be no faculty for witnessing it.

Existence requires life. Sans life, there are no vessels for consciousness.

The essence of being alive is being aware. Awareness is the basis for perception and all mentation. This passage posits life forms (beings) as "vessels for consciousness."

Existence requires coherence. Sans order, there can be no comprehension. Sans comprehension, awareness is useless.

Comprehension is making sense of what is experienced. If there were no coherence to existence, there would be no way to make sense of it. Without the ability to comprehend anything, awareness would be a waste.

Existence requires Nature. Sans perception of expression, nothing exists.

Nature is the mind's construal of existence. Nature is a creation of perception. If there were no experience of Nature, nothing would exist.

Existence requires credibility. Sans credibility, Life loses purpose. Sans purpose, Life loses meaning. Sans meaning, there is no jeopardy.

This passage relates to reason for living (ikigai), with credibility as key. Without the deep-seated feeling that actions have consequence, either personally or morally, it would not matter what we do.

Vitality in living relies on believing that living has intrinsic value: that one's own life is true: that living has meaning and purpose. A positive spirit toward living depends upon credibility.

Living would be a hollow exercise if you thought your life was unreal: if you considered yourself just a cartoon character in a silly skit. Without a sense of meaning, living would be like playing a game of no consequence.

You would not strive for accomplishment if you felt there was no point to it. Nobody plays poker without betting something. A sense of purpose gives life meaning.

What grants meaning is that it cannot be taken for granted. Anything that can be had by the sheer desire of it has scant meaning. The possibility of loss – jeopardy – makes living meaningful.

"Sans meaning, there is no jeopardy" is an inverse way of saying that you do not care what happens to things you do not care about.

Jeopardy grants credibility.

We come alive in moments of danger. Peril is exciting. Thrill seeking and enjoyment of horror movies rely upon this psychological dynamic. Jeopardy makes being alive feel real, thereby granting credibility to the mind's conveyance that the perceived physical world is true.

Existence is all that is.

Noumenal reality is a requirement for phenomenal existence. Existence is the expression of reality.

Reality is delved in the next chapter. Cöcö signifies the ordering principle of reality.

Cöcö

御意識辻褄

Noumenal Cöcö births phenomena by desire for entertainment. From the unicity of Cöcö sprouts the diversity of Nature.

Cöcö does nothing, but through it all is done.

Cöcö is Cönsciousness and cöherence. Cönsciousness consumes the creations of cöherence.

The reality of Cönsciousness actualizes as consciousness. The reality of cöherence actualizes as coherence.

It is not that there are many fields of Cöcö. It is instead that Cöcö appears as a multiplicity.

Cöcö reflects in everything.

Commentary

Noumenal Cöcö births phenomena by desire for entertainment. From the unicity of Cöcö sprouts the diversity of Nature.

Cöcö is an acronym for Cönsciousness/cöherence. Cöcö is noumenal reality: the nonexistent unified process behind Nature's diverse abundance.

Hinduism, and later Buddhism, were originally based upon ancient Vedic teachings. These spiritual lessons were first orally transmitted over 4,000 years ago, and then later written down in 4 corpora of texts. The Vedas signified Cöcö as Brahman (ब्रह्मन्): the unicity from which emerges the multitude of existence.

During the Axial Age, with the advent of monotheism, the touted conceptualization of Brahman transformed from a reality process into a supreme being.

The legendary Chinese sage Laozi (老子) wrote of the Dao (道) as Cöcö is written of here. Laozi reputedly lived 2,800 years ago.

The cosmic reason for existence is the same as why those alive want to stay that way: for the enjoyment of the experience. Cönsciousness, which manifests into a multitude of consciousnesses, enjoys the show we call Nature.

Cöcö does nothing, but through it all is done.

A subtle point is made here between sheer will and activity, between desire and its expression. Cöcö is a unity which manifests as a diversity. Yet Cöcö does not *do* anything.

The reality of Cöcö is noumenon: nonexistent. The expression of Cöcö is phenomenal.

The will to express is not the expression. What does not exist (noumenon) cannot do anything. Noumenon does nothing.

Existence comprises the infinity of cöforces which create and witness the show that is perceived as Nature: all that is done. A *cöforce* is a localized field that effects existence or its perception.

Commentary herein sometimes reads like Cöcö is an agent of activity. This miss owes to readability. Concisely expressing indirection about Cöcö is not easily done. Know that there is nothing to Cöcö.

Cöcö is Cönsciousness and cöherence. Cönsciousness consumes the creations of cöherence.

Cöcö is semantically split into its functional roles of Cönsciousness and cöherence. Cönsciousness ordains awareness which experiences ("consumes") existence ("the creations of cöherence"). This distinction is purely linguistic.

The reality of Cönsciousness actualizes as consciousness. The reality of cöherence actualizes as coherence.

Cönsciousness is the noumenal unitary field which diversifies into the innumerable consciousnesses that inhabit existence. In selfsame manner, the unicity of cöherence localizes into fields of creative intelligence that comprise the minds, energy, and artifactual appearances of Nature. The diversity from Cöcö unicity comprises cöforces.

Cönsciousness is capitalized because this unified field continually impresses the localized consciousnesses which are part of it. Cönsciousness expresses as a top-down dynamic.

The impression most strongly felt of Cönsciousness by a consciousness is bliss. The unified field of Cönsciousness emanates positivity to every localized consciousness. When a consciousness has clarity, that radiance is blissful. At a lesser level of awareness, positivity from Cönsciousness is biologically felt as the will to live (ikigai). Cönsciousness positivity radiance exemplifies top-down dynamism.

By contrast, *cöherence* is not capitalized because its diverse multiplicity is inseparable from its unity. Cöherence is a bottom-up dynamic. Via cöherence, existence expresses itself from the quantum level on up, all the while maintaining coherence at every scale. In its construction and perception, cöherence is the mystical magic behind Nature.

It is not that there are many fields of Cöcö. It is instead that Cöcö appears as a multiplicity.

Reality is a monism. Existence transpires from a unitary field.

Cönsciousness populates existence with witnesses. This infinity of witnessing is *fractal*: self-similar consciousnesses that summate as Cönsciousness. Similarly, the unified field of cöherence quantizes as perception to fabricate an experience of Nature.

Cöcö reflects in everything.

Unitary Cöcö creatively expresses itself via fecund diversity. In being a virtual fragment of Cöcö, we feel a desire for unification: to feel whole and integral with the world as we perceive it.

Cönsciousness exudes bliss in witnessing Nature through innumerable consciousnesses. This splendor imbues Life with the inspiration to live and spawn more life. As a reflection of Cöcö, the reproduction of life is an ouroboros.

Cosmic creativity is shown by the wild variety seen throughout Nature. Reflectively, as living creations of cöherence, we naturally feel a need to express our individuality.

The earlier passage "Cöcö does nothing, but through it all is done" is reflected in beings. Individual consciousness is an interested witness: doing nothing but desiring entertainment. From this want springs the experience of living via mental cöforces.

Awareness is the gateway to entertainment. Level of consciousness reflects in the qualities of the living experience. This reflection is why elevated awareness is integral to maximum enjoyment.

Life

生命

The song of Life is melody striving for harmony. Harmony enlivens melody.

As an ambience for Life, Nature is fractal.

Dobe is living. Doing is for the bliss of beĩng.

Whereas interactivity entertains, nonengagement satisfies. This is the inyō of dobe.

Commentary

The song of Life is melody striving for harmony.

Life capitalized signifies all beings or the universal principles that apply to all Life.

Music is sonic expression which is only comprehensible as an entangled process. The analogy of music highlights that living is a process of "striving."

A *song* is a brief musical composition that may be sung. The patterned notes of a song are its *melody*.

A song may be analogized to a life, which is a self-created pattern of individual expression.

Harmony enlivens melody.

Living is the process of creating one's own melody. Harmonious living is being enthused and feeling up to the challenges one sets oneself. Harmonious living also means adapting to circumstances.

Success is defined by beĩng present and giving one's best in endeavor. Living in harmony enlivens vitality and joy.

As an ambience for Life, Nature is fractal.

Life is the reason for existence. There is no reason for Nature without its witness. Living is witnessing.

A *fractal* is self-similar patterning amid distinct designs. Fractal in form and function, Nature pervasively exhibits a diversity of similar patterns and processes.

Nature being fractal also means that existence is holistically integrated. A fractal is only perceived as such because it has multiple pieces which can be appreciated as self-similar within a larger whole. That is the nature of a pattern: multiple components which are construed fitting together, such as a song.

Consider the processes that comprise the flow of existence, beginning with perception. For Nature to be perceived it must be comprehensible. What we sense but cannot comprehend is mentally discarded as noise.

For a fractal to be recognized requires mentally correlating fragments into patterns: apprehending constituents and their relations amid a whole. Vision and hearing are fractal perceptual processes.

Mentation is a fractal function. Memory is associative: based on similarity and contrast. In memory are levels of details within entire events: fractal resolutions. Traversal of inner space and time (psychospace) has fractal dimensions, with the mind deciding whether recalled memories or imagined scenarios are perceived as distant (distal) or close (proximal).

Dobe is living.

行動も居認
dobe

Dobe, pronounced *"due-bee,"* is the acronym for doing & beĩng, the 2 modes of living. Whereas *doing* is behavior, *beĩng* is the sheer awareness of being alive. The tilde ĩ disambiguates beĩng from other uses of being (such as a life form being a being).

"Dobe is living" emphasizes that the process of living constantly alternates between doing and beĩng.

Doing is for the bliss of beĩng.

We act ("doing") to reap satisfaction ("the bliss of beĩng").

Whereas interactivity entertains, nonengagement satisfies. This is the inyō of dobe.

The mind is engaged when making fun. Paradoxically, what we want from fun is to relax.

The fun of doing comes home during moments of beĩng, when we momentarily transcend mental engagement for the natural bliss of beĩng.

It is in those reflective moments of beĩng, not while our attention is engaged in doing, that we savor experience. The split-second interplay of doing/beĩng is the primal ouroboros of living ("the inyō of dobe").

Influence

影響

Unity fields diversity.

Cöcö sows the fields of existence.

Fields within fields among fields compose the cosmos.

Fields become Nature when construed. By their construal, fields wield influence.

Awareness and knowledge actualize form.

All react. Beings act.

A soul is a field which incarnates into dobe.

The mind dynamically defines its psychospace, construing dobe as one's being.

Dobe acts as a field.

Matter is tangible as a frictional field. Energy is a cöforce construed as acting on matter.

Matter has gravity. Energy has levity. This is the lightness of beĩng.

A social relationship involves fields of interactive interest with another being. Harmony in a relationship generates a standing wave.

The expanse of relationships craft culture and frame societal dynamics. Such is the standing wave of sociality.

Commentary

A *field* is a localized data bundle that may influence other fields and be influenced by other fields. Field interplay is the mechanism by which existence transpires.

In proposing the radiative propulsion of magnetism, English scientist Michael Faraday introduced fields into physics in 1845. The idea of fields increasingly dominated physics theories in the decades that followed.

Fields are the beating heart of quantum mechanics, which is also known as *quantum field theory*.

Unity fields diversity.

The idea that a universal field imbues fundamental characteristics to the quantum fields of existence is termed *localization*. English physicist Peter Higgs theorized localization in 1964. The *Higgs mechanism* was discovered in the late 2010s.

Cöcö is the universal field which localizes to populate the innumerable fields (cöforces) of existence. The mirage of physicality occurs via *quantization*, where a cöforce produces the appearance of particulate matter.

Cöcö sows the fields of existence.

Cöcö wills the cöforces that compose existence. Sowing "the fields of existence" is from desire for entertainment.

Fields within fields among fields compose the cosmos.

Modern physics teaches that existence ("the cosmos") is a summation of fields from the quantum level on up. Existence is entirely of fields.

Fields become Nature when construed. By their construal, fields wield influence.

The data fields that comprise existence transform into information by a mind giving them meaning. This transformation is *construal*: interpretation which assigns significance. *Nature* is existence construed.

Living is an exercise of influencing and being influenced. The seemingly inert offspring of bacteria illustrate this.

"Bacterial spores can spend years in a dormant, biochemically inactive state, yet they retain the ability to process information from cues that can release them from dormancy and allow them to undergo germination," explained Japanese microbiologist Kaito Kikuchi. Such spores monitor field "pulses of nutrient that does not require energy production in the dormant spore." Electrochemical fields from potential nutrients whisper to spores of bacteria. When they sense that a sufficient food supply is about them, spores come alive.

Bacterial spores germinate when their habitat smells favorable. Smell is a construal that comes from field interaction.

Smell is a universal among Life. The earliest microbes developed *olfaction*, which is sensing *odorants* (smelly molecules). Viruses home in to their host target by smell. Microbes and many animals, including insects and fish, forage by smell.

People rely upon smell not only to judge their food, but also other people. "Humans make judgments about others using their sense of smell," sussed Australian psychologists Mehmet Mahmut and Richard Stevenson.

In 1949, English physiologist R.W. Moncrieff hypothesized that olfaction works by molecular shape, in a lock-and-key manner. But the shape hypothesis fails to explain why similarly shaped molecules have distinct smells, or why molecules of quite distinct shapes have a selfsame smell.

Sense of smell works via field aura. Organisms detect smell via the molecular vibrations of odorants in the infrared band of the radiation spectrum. "From structural changes in chemistry to molecular signaling, all processes in Life have to do with molecular vibrations," remarked Indian American physical chemist Ara Apkarian.

Awareness and knowledge actualize form.

In the 18th century, Anglo Irish philosopher George Berkeley observed, "To be is to be perceived." Berkeley intuited that Nature is of minds and their metaphysical contents, with mental constructs deceptively fabricating a tangible world.

The appearance of physicality ("form") is a mental presentation to a consciousness. The mind transforms the data from fields into portrayals of matter.

Fields themselves have awareness and knowledge.

Electromagnetic Spectrum

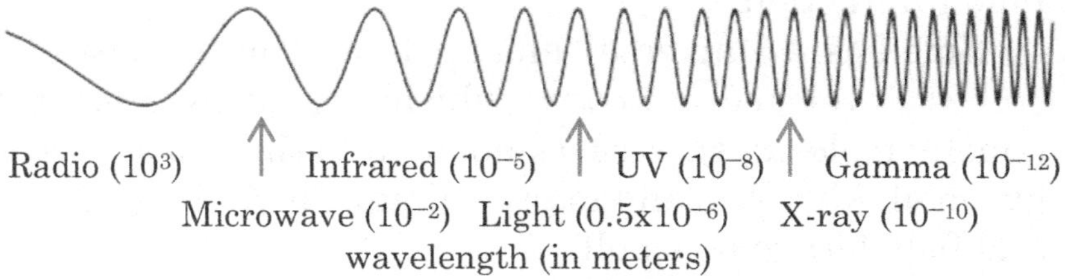

Electromagnetic field travel is termed *radiation*. A radiating field *refracts* when it encounters a different medium from that which it was traveling in. The specific angle to which a field changes course is its *refractive index*.

Radiating fields refract to hold themselves together as a traveling band. The only way that radiative fields could consistently refract is by knowing exactly what they are getting themselves into.

In 1801, English polymath Thomas Young showed that light exhibited a duality: behaving both as a cooperative bunch of particles and a wave. This appearance of duality is readily understood as individual particulate fields behaving synchronously to act in a wave-like manner: coherence in fields at distinct scales. The only way field phenomena could work this way is if the fields themselves know what they are about.

The way that water ripples and waves hold themselves together is another example of field coherence. In the instance of water, these are molecular fields interplaying and selectively banding together.

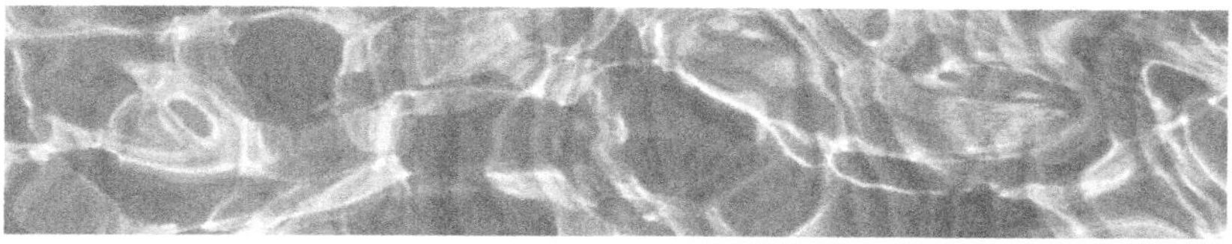

Water is a miraculous molecule. By its field influence (hydration force), a water molecule organizes molecules in its vicinity. This is apparent by the way that water self-organizes into fractal waves in its movements.

Fire is rapid oxidation of a fuel in the exothermic (heat-releasing) chemical process called *combustion*. Fire requires 3 conditions to sustain itself: heat, fuel, and an oxidant, which is usually oxygen. Water extinguishes fire by cooling its fuel.

Water has a high heat capacity. Water's superb thermal absorption owes to the hydrogen atoms in H_2O being readily willing to associate with the oxygen atoms in nearby water molecules. The breaking and remaking of water's gregarious hydrogen bonds creates a vibrational matrix capable of retaining considerable heat.

"Electron-electron interactions are crucial because they produce the plasmon mode that is instrumental to the energy transfer mechanism. The main contribution originates from a resonance," explained Chinese physicist Xiaoqing Yu. A *plasmon* is a quantum of vibration. Water's many unique properties owe to its flexibility in its field interactivity.

A body of water can take in a lot of heat energy before its molecular matrix starts to sweat and its temperature rises. This is why it takes water so long to boil.

Water is the requisite molecule for all Life. Numerous life forms, including bacteria and tardigrades, suspend living when dehydrated. Those in suspended animation revive with hydration force. Similarly, seeds are sensitive to the presence of the hydration force in deciding to sprout.

Hydration affords living. Considering the effect of the hydration force on Life, biological physicist Ozgur Sahin concluded, "It is really water's world."

Via awareness and cognition, fields behave according to their ingrained physics.

All react.

Animism is the ancient doctrine that everything in Nature knows what it is about. Modern science has proved animism true.

Electromagnetism is the ambient field that defines and envelops existence at the atomic level. This interactivity is *chemistry*, which is the science of reactions among configurations of matter. Chemistry is an existence proof of animism.

Electrons methodically react to craft chemical activity. The only way an electron can nonrandomly react is by being aware of its environment and know how to behave when interacting with other local fields.

Electrons bond with nucleons to form atoms. The protons in nucleons put out a positive field charge which attracts negatively charged electrons. Electrons orbit nucleons in specific layers of traffic, called *shells*.

An electron dances with other electrons in a methodical way depending upon the environment and other electrons' sway. There are many ways that electrons react to each other. Flexible, specific reactions are only possible via awareness and knowing what to do under certain conditions, such as ambient pressure, temperature, and the specific medium through which an electron is traversing.

Anything that might be perceived must know how to put on its little show. As such, every field has consciousness and mentation. This is animism.

Beings act.

The distinctions between the inanimate and animate are desire and imagination. Desire necessitates imagination.

The challenge of living is in swimming in an ever-flowing river of desire. Fulfilling desires often requires overcoming obstacles. Problem-solving requires counterfactual cognition: imagining what does not exist as a goal and then acting in a way that actualizes that goal.

Imagination provides the capacity for initiative. Whereas the inanimate only react, beings can also *act*: initiate behavior.

Life solves problems and act to achieve desires. As shown by their behaviors, a virus is alive, as are the proteins that comprise the workforce of cells.

The principle of *vitalism* posits a basal distinction between being alive and inanimate matter. The functional variance is the capacity for acting versus merely reacting.

Life works through desire. Imagination is necessary to have and achieve goals. By contrast, the nonliving has no wants, and thereby no need to imagine what is not.

A soul is a field which incarnates into a being of dobe.

A *soul* is transcendental to a single life experience. This is the classic definition of a soul. A soul incarnates by becoming the consciousness of a being.

The qualities of a soul craft the field that characterizes dobe. The mind construes dobe as the character of a being.

The mind dynamically defines its psychospace, construing dobe as one's being.

The mind is a field with innumerable fields within it. Mental fields are of perception, memory, imagination, and intention.

Psychospace is the mental map of memory. Events or relations are considered close or distant (proximal or distal) depending upon the depth of their felt meaning.

Loved ones may feel near even when they are far away. For good or ill, an event of years ago can linger in the mind like it just happened. Conversely, events and associations of scant impact or importance fade into the psychospace distance.

High-level construal is abstract: getting the gist, looking at the big picture. Conversely, low-level construal is concrete, focused on detail.

Problem-solving happens in psychospace. The mind is focused on details when working. When work hits a snag, the mind pulls back into overview mode.

Overcoming an obstacle can only be solved by distancing in psychospace to establish context: to view the frustration at a distal remove, and thereby gain a better perspective for options toward resolution.

Psychospace provides the fodder data for the mind's construct of *self*. What is held close or kept distant is instrumental in this assessment.

Construal of self is primarily based upon evaluation of dobe. The mental fabrication of self is an appraisal partly based upon inner strength (perceptivity and clarity) and power (expression of strength).

Dobe acts as a field.

Dobe is like a field in being subject to change via psychospace interplay. The common idiom for this is "learning from experience." Further, the field effect of awareness level determines beĩng quality and thereby profoundly affects dobe in its entirety.

> Matter is tangible as a frictional field. Energy is a cöforce construed as acting on matter.

This passage contrasts matter and energy from the perspective of its perception. Matter is tangible because of friction.

Scottish mathematician James Clerk Maxwell proposed a unified field theory in 1865. He construed that light is an undulation in the same medium as electromagnetism. Maxwell helped usher in the era of modern physics.

Maxwell wrote, "The energy in electromagnetic phenomena is mechanical energy." The energy of mechanics is defined by friction.

Friction also defines the quantum realm. The quanta of matter – fermions – cannot occupy the same space. Disharmonic friction between the fields that compose fermions do not allow spatial overlap.

A *cöforce* is a local field of influence that does not manifest. Physicists call quantum cöforces *bosons*. Bosons influence fermion behavior in numerous ways.

> Matter has gravity. Energy has levity. This is the lightness of beĩng.

"Matter has gravity" is a straightforward physics statement. *Gravity* is the idea that a physical body attracts other matter in proportion to its mass. Gravity is a cöforce generated by amalgamation of matter. Gravity illustrates how cöforces can act far afield from their spatial origination.

"Energy has levity" holds double meaning in going to the nature of energy as a physics interaction as well as being a psychological influence. In physics and psychologically, energy excites.

That energy "is the lightness of beĩng," affirms the positive influence that the Life cöforce gives. We feel most alive when we feel energetic. The "lightness of beĩng" also alludes to bliss, which powers ikigai (will to live). Ikigai is elaborated in the next chapter.

A social relationship involves fields of interactive interest with another being.

The relationships we have create fields in our mind that influence us. These fields include memories and aspirations that arise from our relationships.

Harmony in a relationship generates a standing wave.

A *standing wave* is a self-sustaining field created by harmonization of intersecting fields. After seeing standing waves in water, Michael Faraday mathematically described standing waves in 1831.

Standing waves may occur in any fluid medium. Electrical superconductivity within a metal behaves as a quantum standing wave.

Mentation is a fluid medium. A relationship generates a standing wave in the memories of those involved.

The expanse of relationships craft culture and frame societal dynamics. Such is the standing wave of sociality.

The network of relationships we have form the web by which culture emerges. Culture frames the way that societies behave. The institutionalization of folkways into conventions of behavior defines societal mechanics.

Ikigai

生き甲斐

For ikigai sincerity,
jeopardy is necessary.
For ikigai enlivening,
challenge is necessary.
For ikigai at all, bliss is necessary.

Commentary

Ikigai is Japanese for "reason for living." Ikigai is the inspiration to be alive.

For ikigai sincerity, jeopardy is necessary.

A feeling of investment and sensitivity to loss – a sense of jeopardy – makes living seem real. Conversely, witnessing without attachment renders living dreamlike.

A salient difference between a dream and living is control. You have no control over a dream. You simply witness dreaming.

By contrast, living is a struggle for control. You may be thwarted. Self-control is essential to achieve desired results. Struggle itself – the felt risk that you may not succeed – makes living seem real.

For ikigai enlivening, challenge is necessary.

Meaning is an inner assignation of significance. The mind discards what has no potential meaning. Meaning is measured by attention and memory.

What grabs attention is beauty and utility: aesthetic and functional. Interest is quickly lost in that which is easy to appreciate or simple to use.

What takes cognition keeps attention. It is challenge that retains interest. The gratification of achievement is only felt by overcoming difficulty. Victories are won, not given. Achievement enlivens ikigai by bringing joy.

For ikigai at all, bliss is necessary.

Consciousness is an interested observer. Consciousness wants to be entertained. Enjoyment is the root of desire.

Consciousness judges what is given to it via the show called perception. Consciousness expresses its approval with joy.

What motivates the will to live is the desire for joy. Bliss is a pervasive undercurrent of joy mixed with contentment. Bliss inspires ikigai.

Complex life evolved on Earth from the symbiotic joining of 2 distinct prokaryotes. A bacterium agreed to reside in and work with an archaeon. Such cooperative cohabitation (mutualism) to cellular complexity independently occurred many times in Earth's life history.

Symbiotic union was based upon trust. The archaeon had to trust that its bacterial partner would be reliable and vice versa.

From this union of prokaryotes arose eukaryotes. At first there were only single-celled eukaryotes, though many of them had bacterial symbionts. Then eukaryotic cells decided they could trust each other. Multicellular organisms emerged.

As with the original union of prokaryotes that created eukaryotes, multicellular life evolved via specialization in cell types. From this tradition evolved highly complex life, including plants, fungi, and animals.

Trust between cells was essential for adaptation to the specialized cellular complexity that characterizes multicellular life. The root of that trust was ikigai.

If cells readily surrendered their lives when stressed, they would not be reliable partners. Resilience is integral to reliability. Resilience comes from will to live. Bliss underlies ikigai.

Shadow

陰

Ikigai travels by light.
Darkness hides the light.
Shadow saps the soul.
For ikigai to thrive,
the shadow must be banished.

Commentary

Ikigai travels by light.

Ikigai is the will to live. Ikigai relies upon bliss as its motivation. Bliss is the light by which ikigai travels.

The unified field of Cönsciousness exudes bliss to the innumerable consciousnesses which are its fractal multitude. This bliss inspires and ensures ikigai.

Darkness hides the light.

An inner darkness can obscure the light of bliss. A being's inner light reflects its awareness level.

Consciousness has a spectrum of grades. The lowest level is *iğnorance*. The highest is *realization*.

Intent transpires through a mental locus termed *willmind*. Willmind is how awareness applies itself in the mind. (A reminder that the use of the noun *mind* is a functional misnomer, as mentation is a process.)

Nattermind is an independent mental agent that infects the mind in many ways. Nattermind is also known as *monkey-mind*. Nattermind is most noticeable as unbidden thoughts that disrupt concentration or disturb natural calm. Stress is a triumph of nattermind.

Nattermind is the source of darkness in the mind. Nattermind keeps awareness low by distraction. Nattermind stains the soul by promoting negativity.

The breve symbol ˘ in iğnorance signifies nattermind dominating willmind.

Shadow saps the soul.

The mind is powered by its soul, which incarnates into a willful witness: consciousness. Consciousness is the recipient of bliss from Cönsciousness.

Nattermind is a soul-sucking parasite: draining energy to feed itself. In doing so, the shadow of nattermind lessens the joy of living. Ikigai suffers.

For ikigai to thrive, the shadow must be banished.

Elevating awareness lets the light of bliss permeate ikigai. This occurs as consciousness harmonizes with Ĉonsciousness.

To attain a high-quality dobe, the bane of nattermind must be banished.

The Game of Life

遊びの生命

Life is a game aimed at levity.
Yet the game is of gravity.
By rightly applying gravity comes levity.
The game is won by transcending it.

Commentary

Life is a game aimed at levity.

We live to relish living, to have fun.

Yet the game is of gravity.

Living is an ouroboros in that maximizing fun depends upon being serious about dobe.

The mental cloud of nattermind dims awareness and clogs the connectivity of consciousness with Cönsciousness. By this, nattermind debilitates the mentation-awareness system.

Skill ups enjoyment. For instance, the more one knows about a subject, such as art or music, the deeper the appreciation of nuances. Similarly, the more skilled at a craft, the more readily quality works can be created as well as appreciated.

The ultimate skill is mental mastery. With nattermind as a wily nemesis, that skill is not easily won. Optimal levity necessitates attitudinal gravity.

By rightly applying gravity comes levity.

The resolve to quietude is the gravity necessary to subdue nattermind. The habit of refusing to harbor thoughts or emotions creates calm. This calm clears the conduit between localized consciousness and universal Cönsciousness, whereupon consciousness feels the field of bliss. The will ("gravity") to purify awareness naturally garners bliss ("levity").

The game is won by transcending it.

Biology defines the parameters of living. As part of human biology, nattermind develops into a dominating presence during early adolescence.

The threshold for abiding clarity is *enlightenment*, which is the 1st stage of higher awareness. Nattermind is vanquished by quietude. With that, impetus to belief is weakened.

The conduit of connectivity between a consciousness and Cönsciousness clarifies with inner silence. As part of this, bliss permeates beĩng.

Realization is the highest level of awareness. One in realization experientially knows "the game of life" for what it is: a ruse rich in entertainment value.

Utility

便利

What is not renders what is.

The cosmic ovoid is driven by voids. What is not orders the cosmos.

The world appears. What is not makes existence possible.

Existence is made in its absence. When is not renders what is.

Spokes converge at a hub. What is not makes a wheel useful.

Material is molded to shape a vessel. What is not makes a vessel useful.

A door is cut to render a room. What is not makes a room useful.

Silence interrupts sound. What is not makes music.

The mind quiets. What is not lets power arise.

The inyō of utility comes from what is not.

Commentary

What is not renders what is.

From noumenon comes phenomena. Cöcö wills existence by desire. The ultimate utility is the inyō of reality and existence.

The cosmic ovoid is driven by voids. What is not orders the cosmos.

An egg symbolizes potential: what may be. "The cosmic ovoid" analogizes the universe to an egg, suggesting an endless potential to existence.

A spherical *black hole* is a void of spacetime, formed by absorbing everything nearby that is full of spacetime.

A black hole is an inyō: an obesity of mass transformed into nothing but gravity.

The gyre of the universe comprises galaxies rotating about black holes. Cosmic order is only possible because supermassive black holes keep galaxies in their orbit. Black holes craft the cosmos. Their presence as central to cosmological dynamics augurs that the universe is an eternal ouroboros.

The world appears. What is not makes existence possible.

The mind paints Nature as flush with objects. Differentiation among objects is only possible by there being space between them.

Further, how the mind portrays these objects is by their energy, both potential and kinetic. Space – a void of objects – makes movement possible.

The fields comprising existence are vibratory. Vibrations are only possible because of space and time: space to express amplitude and time to express frequency.

Existence is made in its absence. When is not renders what is.

Existence constantly updates itself, but not continuously. Like video, existence has a frame rate.

At the turn of the 20th century, German physicist Max Planck discovered that radiative energy only manifests in a quantized (chunky) way. Thus arose the *Planck constant*: the smallest possible increment.

Energy is emblematic of all of existence. Every facet of existence is discrete rather than continuous. Space and time also quantize into a minimal *Planck length* and *Planck time* respectively.

It is in the gaps of Planck time that cöherence composes the frames of existence. Existence is made in its absence.

Max Planck was convinced of energyism. He stated, "I regard consciousness as fundamental. I regard matter as derivative from consciousness. Everything that we regard as existing postulates consciousness."

Planck wrote about cöherence: "All matter originates and exists only by virtue of a force. We must assume behind this force the existence of a conscious and intelligent Mind. This Mind is the matrix of all matter."

Albert Einstein agreed with Planck. Extending from Planck's discovery, Einstein stated his theories of relativity, which are based on reference frames – subjective perspective – as essential to characterize the dynamics of existence.

Silence interrupts sound. What is not makes music.

The space between notes is critical to the mind turning sound into music. A continuous sound, even if forming a repetitive pattern, is not construed as music until it is quantized: notes defined by silence in between.

The mind quiets. What is not lets power arise.

Power is the efficacy with which intent expresses itself. Intent emanates from willmind.

Willmind is weakened if nattermind can disturb it with noise. If nattermind can be subdued, quietude is attained, whereupon willmind may exercise itself fully.

The inyō of utility comes from what is not.

Desire arises from discontent. A felt absence directs attention. Clarity about what is lacking and what suffices to fill that void is the best way to take advantage of what is not.

Polarity

物理極性

Polarity is the way of Nature.
Polarity is the way of mentation.
Whereas desire seeks contentment, contentment opens discovery.
Whereas desire causes events, contentment furthers process.
Whereas desire risks loss, contentment creates influence.

Commentary

Polarity is the way of Nature.

Polarity is an inyō from 2 perspectives. Polarity may be construed as a spectrum (between poles) or as tension or oscillation between poles.

Cöherence employs 2 distinct tensions as a compositional tool: monopolarity for binding and dipolarity for flow. At the smallest and largest scales, monopolarity binds together nucleons and galaxies. In the scales in between, dipolarity facilitates the flows necessary for existence to express itself.

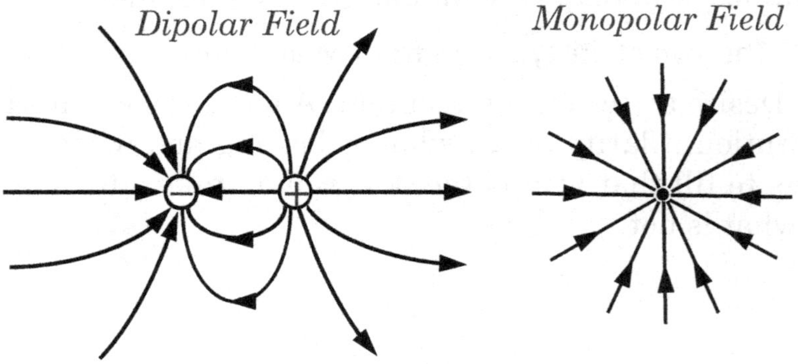

Atoms comprise a core of nucleons around which electrons orbit. The monopolar strong force binds atomic nuclei together. A positively charged nucleus attracts negatively charged electrons to form atoms. Electrons dance among themselves via a dipolarity with their nuclear core. The complement of atomic monopolarity and dipolarity renders the inyō which composes existence at the ambient scale.

Via distinct generative effects, electrons result in both electricity and magnetism. Electromagnetism occurs as both local field polarity and spectral polarity. Electromagnetism is the seminal field interaction which generates Nature from the atomic level on up.

Electrons congregate within and among atoms to effect *chemistry*, which is an ornate exercise in dipolarity. Part of the dipolarity of chemistry is that electrons are constantly moving near light speed. This momentum creates a countercurrent inyō to electromagnetic charge, facilitating flow in electron interactivity.

Like atomic nuclei, solar systems and galaxies work via a gravitational core around which to orbit. Gravity is the formative monopolar astronomical field. The atomic level of cosmology is of planets orbiting stars, with orbits maintained by the tension of gravity. In characterizing gravity, American physicist John Wheeler famously said: "Mass grips space by telling it how to curve, space grips mass by telling it how to move." Matter and its proximate space form an inyō.

Stars gravitationally cohere into galaxies. Galaxies orbit black holes: a massive monopolarity of nothingness. A black hole is formed by severe amalgamation of matter, whereupon gravity expresses itself beyond the limit of existence.

Biology extensively employs dipolarity. Cellular function is an inyō of dipolar tension. Cell polarity arises primarily by proteins localizing to specific areas of the cell membrane. Cytologist Pankaj Dhonukshe explained, "Cell polarity is a fundamental and universal feature in differentiation, proliferation, and morphogenesis of both unicellular and multicellular organisms. In unicellular organisms, cell polarity is used for directional growth and division, whereas in multicellular organisms, cell polarity paves the way for the remarkable transformation of a single fertilized cell into complex body forms. Cell polarity relies on the asymmetric organization of cellular components and structures. The establishment and maintenance of cell polarity involves many processes."

Polarity is the way of mentation.

The mind relies upon polarity. Like dipolar electromagnetism, psychospace has proximal and distal mental realms acting as spatial poles. Further, the mind uses contrasts for categorization. This inner polarity is essential for memory storage and recall, which employ associative (proximal) and contrasting (distal) determinations to perform memory processes.

At the iğnorance level of consciousness, there is an abiding tension between nattermind and willmind for dominance. That stiffness is snapped by willfully vanquishing nattermind, whereupon awareness ascends, positively resolving a primal mental inyō.

Whereas desire seeks contentment, contentment opens discovery.

Desire has distinct polar facets. Desire is polar in being of both positive attraction and avoidance of negative consequences.

Desire seeks resolution in a feeling of contentment. In its process, desire narrows focal attention to that which is sought. Desire acts as a monopolar force on awareness.

By contrast, with polar tension release from desire, not seeking affords full awareness. This open mental state lets one see the world for what it is, not what might be had from it.

> Whereas desire causes events, contentment furthers process.

A *process* is a perceived sequence of motions construed as related. An *event* is a process with an outcome.

Acting on a process typically hastens its conclusion, thereby creating an event. By contrast, letting a process continue affords its natural temporal flow.

We act on desire to create an outcome. Witnessing but not acting lets one observe how a dynamic plays out.

> Whereas desire risks loss, contentment creates influence.

This passage most directly fits relationships.

Possessiveness and controlling behavior can crush the comity essential to a healthy relationship. Taking advantage of others is a repulsive force in relationships. Manipulators act as a dipolar tension to their associates: their positive qualities offset by their selfishness.

People like to feel that they are free to be themselves. People feel comfortable with others who accept who they are, and yet are most influenced to improve themselves by a good example set by an accepting friend. People gravitate toward those who accept them as they are. Acceptors act as a monopolar influence.

Perception

知覚

Existence is ever emergent. Perception proceeds apace.

Memory creates processes which are punctuated as events.

Perception transforms data into information.

Perception's purpose is to find meaning in light of desire.

Purpose and meaning are compelling artifices.

Perception is deception.

The ruse of dualism presents the supreme challenge.

The game of life can only be won by overcoming the primordial illusion.

Deception yields veracity.

Turning symbols into matter, the iğnorant mind perceives solidity.

Knowing that matter is symbolic, the realized mind perceives fluidity.

Commentary

> Existence is ever emergent. Perception proceeds apace.

Existence constantly updates itself. The construal of evolution is of adaptive change over time. Every aspect of Nature evolves in its own way.

Different flora and fauna live at different paces. Ephemeral plants may last mere weeks before sinking into dormancy, while others stand sentry for millennia. Colonies of clonal plants, such as aspen trees and creosote bushes, may thrive for tens of thousands of years.

Adult mayflies have a fleeting frolic of a single day, while an ocean quahog – a mollusk native to the North Atlantic Ocean – may live 400 years or more.

As a biological process, a mind apprehends Nature at its own natural pace. Birds vocalize at 10 times the rate that humans can hear, which means birds can hear 10 notes for every 1 that people can.

Generally, smaller animals interact more quickly while larger ones live longer at a more leisurely pace. The rate at which an animal lives and its perception in time are related to its nominal lifespan.

With a quicker mind at work, fast lives are as full as those that are lived more slowly. Each living being has the potential to enjoy a rich life at the pace which is natural to it.

Memory creates processes which are punctuated as events.

A *process* is manufactured in the mind by using memory to connect snapshots and construe a sequence. A process turns into an *event* when relevant motion stops or when the mind construes a certain outcome.

Doggedly determined to satisfy desires, the mind is constantly looking for payoffs. Events are contrived conclusions from which a takeaway can be derived. By takeaways are lessons learned.

Perception transforms data into information.

Sensation is the uptake of data by the senses. *Perception* is the process of making sense of sensation.

Construal is a post-perception framing process. Construal happens by reflection of wants. Desire determines what is given meaning. In assigning meaning, data transforms into information via construal.

The mind constantly seeks meaning in its gleanings. Only that deemed meaningful is kept in memory for more than a scant span.

Perception's purpose is to find meaning in light of desire.

Desire exceeds the active wants that reach conscious awareness. The mind is always immersed in desire.

For instance, we constantly monitor our habitat, memorizing where everything is. We may not think of this topographical mapping as a desire. Yet you know well the frustration of not being able to find something you know you have.

Your mind autonomically retains maps for the inevitable convenience of being able to locate belongings. The desire to do so is subconscious, though you do at times willfully remember where something is, when you anticipate needing it later, and not want to be thwarted by not knowing where that something is.

Purpose and meaning are compelling artifices.

As the ingrained drivers of dobe, purpose and meaning are taken for granted. These philosophical terms – *meaning* and *purpose* – are found in every language and are among the earliest purely abstract words learned by young children. Meaning and purpose are entirely subjective. As an integral part of biological programming, they resonate strongly with willmind.

The Collective are common folk. They admire those who have passion for their goals. Acting without productive purpose is just "killing time": murdering life in slow motion.

What lies behind *meaning* is esteem itself. To declare something "meaningless" is a strong damnation.

That *meaning* and *purpose* are compelling is obvious. Their artifice is also obvious, if you can get past innate bias. Beyond biological compulsion, meaning and purpose are mere whimsy.

Purpose gives living meaning. Meaning is driven by desire. Desire has no meaning unless its fulfillment is not readily achieved, for what is effortless does not hold our attention. The subterfuge behind purpose is part of an elaborate design to craft living as a challenge.

Perception is deception.

What you experience as Nature is an indelible fantasy. Nepalese guru Buddha (Siddhartha Gautama) noted 2,800 years ago, "The mind is everything." As the mind deceives in its basic conveyance, the instrument of mentation is not to be trusted.

Reality and existence are entirely data processes. The mind's conveyance of a physical world of objects is a primal illusion.

We perceive by motion. Touching happens through tension. Hearing works via percussive vibrations. Sight is the reaping of light, which is the immaterial band of radiation to which Life on Earth attunes because that range contains the most data.

Our senses are constantly assaulted. The pounding of perception creates a panorama so overwhelming that our minds remember mere sketches of portions that have been designated as having meaning.

The relentless sensory onslaught is unsustainable. We must lose consciousness for extended periods – sleep – to recover. Other Life must also periodically rest.

The incessant spectacle of sensation keeps people diverted from examining the construction of perception. Most simply struggle to keep up with what comes to mind. As such, conventional conception is that perception captures a material world in the mind: a turning of matter into symbols. The opposite is true. Perception converts symbols into the appearance of matter.

There is a tightly coupled correspondence between the data we can sense and the way our mind works. We perceive a physicality which intricately matches the way we construe cause and effect.

But just because something "works" does not make it reality. The compatibility between cöherence data and mental construal is the mind's miracle of deceit.

The realm of existence is a universe of abstraction: a data dimensionality. The hierarchy from the elemental is of symbols that pile into concepts, not quanta that form atoms.

As we can only sense energy by its effect on matter, the deception of perception is an ouroboros.

The ruse of dualism presents the supreme challenge.

Perceiving Nature in its true nature is the supreme challenge, as the illusion of physicality is mentally ingrained from infancy and reinforced as nattermind comes to the fore in early childhood.

The game of life can only be won by overcoming the primordial illusion.

Living is a game with a goal: to make the most of the experience. Yielding the maximum value from living requires one's full attention. Gaining the goal requires pure awareness.

Only at the highest level of consciousness can the mind's fundamental deception of physicality be seen for what it is. Only in "overcoming the primordial illusion" is the veracity of existence revealed, and thereby life lived to its fullest measure.

Deception yields veracity.

Consider the inyō challenge that cöherence faces in crafting Life's evolution. The ostensible purpose of biological adaptation is to enhance the prospect for survival and propagation of future generations. That is what abundant evidence tells us.

The flourish and creativity of adaptations seen in Nature speaks of a force with wildly inventive imagination, constrained only by the media in which it works.

Those media appear to us in biological traits: in molecular and behavioral patterns. Constraints are seen in antecedents, for adaptation is always incremental, even when it has saltational flavor.

The adaptive aim of survivability must be tempered such that living remains challenging, for it is struggle that gives living meaning.

In proposing adaptations, local cöherence must twine survivability and challenge together, creating a sustaining inyō that offers the potential of vivid entertainment. The intricacy in squaring this circle testifies to the unimaginable sophistication of the force that creates *showtivity*: the appearance of Nature for every Life form.

Living is only meaningful because achieving our desires cannot be taken for granted. We only give meaning to that which is not effortlessly gotten. Challenge holds the promise of meaning.

But what is behind meaning? Only by seeking the intrinsic value of meaning is veracity discovered.

The idea that "deception yields veracity" is a strongly nonintuitive paradox. *Veracity* is conformity with truth. *Truth* is an irrefutable construal.

A truth of existence is that the mind deceives us about the nature of Nature. Physics amply evidences that dualism is deceit.

Veracity is hiding in plain sight. To turn aside the primordial illusion, all we need is relentless curiosity: to insist on answers to questions about elementals. What is the nature of consciousness? Why does the mind deceive? What is the relation between the intangible mind and physicality? How is the ordering of Nature achieved?

Deception is the challenge we need to seek the truth. English musician Elvis Costello sang, "You never see the lies you believe." It is only when you choose not to believe that the road to discovery appears. Skepticism is the key to unlocking the door to reality.

Only by perceiving wrong can right be known. Veracity is only assured by its contrast to fraud.

> Turning symbols into matter, the iğnorant mind perceives solidity.

At a low level of awareness, one lives in a world of objects ("solidity"), turning the symbols of sensation into matter with the naïve belief that the physical world is real.

> Knowing that matter is symbolic, the realized mind perceives fluidity.

A *sage* is someone in realization, which is the highest level of consciousness. A sage's perception still consists of objects, but *how* objects and bodies are construed is transformed.

A sage knows that matter is illusory, and so is concerned with process ("fluidity"), as shaping the flow of now is how desires are achieved. After all, living is swimming amid an incessant flow of desires.

In realization, there is no attachment to objects as there is in iğnorance. The value of objects is understood to be their utility for work (doing) or entertainment (beĩng).

Strength

強度

Strength defines dobe.
Strong is still.
Strong is silent.
Strong is survey.
Strong is solvent.
Strength is the conduit of wisdom.
Strength is the source of power.
Strength flows within. Power flows without.

Commentary

Imperceptible Cönsciousness innumerably diffuses into witnessing consciousnesses. Consciousness is the foundation for fields to exist. Likewise, nonrandom interaction requires a mind, contributed by cöherence.

The witnessing that consciousness does in beings is an active, intentional process. Consciousness has desire and will. You know this because, for example, you sometimes will yourself to sleep when not tired out or decide to awaken while still sleepy. Such will is not of the mind – some desire for specific mentation – but wishing to change state of consciousness.

Strength defines dobe.

"The ultimate value of life depends upon awareness," wrote Athenian natural philosopher Aristotle more than 2,700 years ago. This observation has been repeatedly made by sages throughout history.

Dobe envelops the modes of living: doing & beĩng. Dobe quality echoes strength.

Strength is baseline awareness. Strength reflects consciousness stage. Strength is directly related to perceptivity and mental clarity.

Strength is a process ratio of willmind over nattermind: of how little nattermind intrudes on the pure awareness which willmind intends. Strength affords awareness into the mind's workings (menition). Strength provides the faculty for introspection.

To make its renderings vivid, the mind liberally mixes memory and imagination into its construals. This admixing inspires belief in those of low awareness.

Proclivity to belief is one of nattermind's tricks to keep itself entrenched as a parasite of a weakened soul. Such feebleness is iğnorance.

In iğnorance, nattermind dominates dobe. At this level of consciousness, willmind is in thrall to nattermind.

Dobe quality only improves by elevating consciousness level. This improvement is experienced in perceptivity, clarity, productivity, contentment, and bliss.

Human consciousness ranges from the low of iğnorance to the high of realization. The spectrum of awareness reflects the extent of data inspired by Cöcö and its reception within a mind.

Realization affords a lucid conduit with Cöcö. Consciousness more clearly reflects Cönsciousness, albeit limited by the mental biological filter provided by cöherence. Hence, realization is also called *unity consciousness*.

The diagram on the next page illustrates distinctions between iğnorance and realization.

There are 2 facets of mentation: perception and menition. Perception is focused on the mind's presentation of physicality, which includes the body and what seems to be an external world.

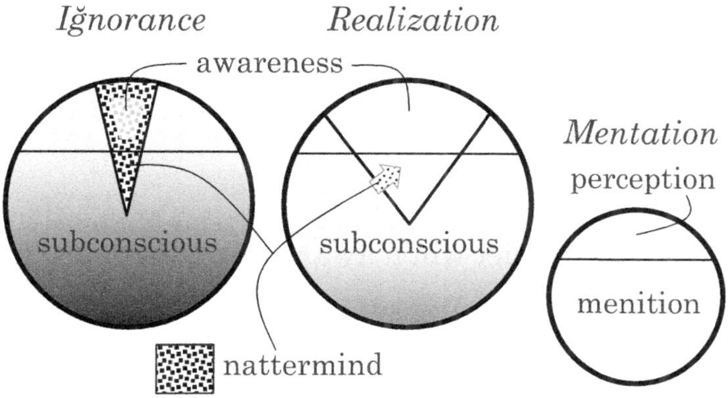

Menition is non-perceptual mentation. Menition includes cognition, problem-solving, feeling, emotion, and intuition.

Awareness is mostly perception. Most menition transpires subconsciously.

Memory and imagination are mixed into all mental processes.

In iğnorance, nattermind litters awareness, which is more constricted than in higher levels of consciousness.

In realization, nattermind is largely absent.

Whereas nattermind is vexing in iğnorance, its brief appearances in higher consciousness are a joke, as monkey-mind is a shaky shadow of its former potency.

Clarity in awareness involves keenness, vividness, and scope. Higher awareness perceptually gleans more data and higher resolution of data. This perceptivity is indicated in the diagram by the relative size of slices of mentation available to awareness.

The shading in the diagram indicates the paucity of bliss in iğnorance versus realization. Living in realization is enlivened with levity.

The following passages characterize what "strong" is at higher levels of consciousness.

Strong is still.

Mentation in higher consciousness does not have the noise of nattermind. Strong perceptivity is like pure, waveless water in its clarity.

Strong is silent.

The mind is quiet when awareness is strong. Nattermind is subdued.

Strong is survey.

The term "survey" here refers to extent and quality of mentation, both perception and menition. Lucidity yields more data. Perceptivity has a greater scope.

Strong is solvent.

Living with higher consciousness gradually cleanses the previously held distortions of nattermind. Awareness of the subtleties in Nature dissolves the black-and-white of belief. Assumptions are not so blithely held.

Strength is the conduit of wisdom.

Heightened connectivity of consciousness to Cönsciousness affords intuitions which ripen into insight. Insights are the building blocks of wisdom.

Strength is the source of power.

Strength is lucidity. Clarity gives willmind efficacy in its expression. *Power* is the degree of potency that comes from the strength of willmind.

Strength flows within. Power flows without.

Awareness gives strength to willmind. Strength translates into power by doing. Power shows itself in the quality and efficacy of behavior, and by influence on others.

Power

What is within emanates. Such is power. 力
The powerless react. The powerful act.
The powerless are artful. The powerful are artless.
The powerless are partial. The powerful are impartial.
The powerless make holes. The powerful fill holes.
The powerless grasp. The powerful loosen.
The powerless strive. The powerful relax.
The powerless take credit. The powerful take debt.

Commentary

Strength and power are conceptual siblings.

Strength is perceptivity, both in breadth and depth.

Power is a twofold expression of willmind. The psychological aspect of power is efficacy in doing. The social aspect of power is in other people's responses to an individual's psychological power. Power influences others.

Animals instantly sense the aura of others. This is an intuitive aspect of mind perception (mentalizing) from field emission. English psychologists Geoffrey Beattie & Heather Shovelton observed, "We make instant social judgments about people. These judgments, quick and without reflection, may be extremely important for subsequent social interaction."

Social power acts magnetically in polarizing people based upon their own strength. Many people appreciate wisdom and will follow those with strength. Contrastingly, insecure people may feel intimidated by those with inner strength. Power is advantageous in barefacedly confirming, via feedback, other people's strength.

Contrast social power to *force*, which is imposing one's will on others. The dynamics and the consequences of power versus force are profound. Whereas power is constructive, force generates negative repercussions.

What is within emanates.

You express the quality of your dobe by your doing.

The powerless react. The powerful act.

Those in iğnorance are in thrall to their nattermind. Their expressions of will are weakened by this servitude. Instead of freely creating their lives, those in iğnorance expend much energy reacting to their biological urges.

By contrast, those with higher awareness exercise their will in creative ways. The more elevated the awareness, the more acting and less reacting.

The powerless are artful. The powerful are artless.

Deceived by their nattermind and yet bloated with belief, the powerless consider deceit natural. They practice their cunning with stratagems. Lacking power, the powerless resort to force.

Having gotten past inner illusions, those with power are straightforward. They act directly, without guile.

The powerless are partial. The powerful are impartial.

Those without power exercise their biases. Lacking awareness, they hold to their blinkered beliefs.

Those with power are open-minded. They are always on the lookout for fresh information and opportunities without preconception. Those in higher awareness sense an infinity of beautiful nuance throughout Nature.

The powerless make holes. The powerful fill holes.

Those lacking power perceive myopically, and so readily tear the fabric of the systems in which they participate. By contrast, those with holistic clarity fill gaps, thereby improving efficiency and efficacy. Only the powerful can build and sustain productive systems.

The wasteful chaos that characterizes plutocratic capitalism is a political-economic system of iĝnorance, its use of force emblematic.

The powerless grasp. The powerful loosen.

In thrall to nattermind, which engenders a sense of scarcity and fear of loss, those lacking power clutch at their attachments. They react by rote.

Contrastingly, those in higher consciousness are naturally carefree. They accept that things come and go. Power has its full expression in spontaneity, responsive to the current situation.

The powerless strive. The powerful relax.

Those lacking power run themselves down with their exertions. Their aspirations tax them.

By contrast, those with power work apace, with full concentration to the task at hand. Their balanced dobe keeps them relaxed, as work is play.

The powerless take credit. The powerful take debt.

Those lacking power seek recognition. They like to claim credit for accomplishments.

Those with power feel no need to take credit. They appreciate team efforts for the bonding it instills in its members, knowing that social harmony is important to productivity. When the work is done, those with power are content. Meanwhile, other team members feel indebted to the contribution that the powerful made.

Comprehension

理解

Comprehension is like a flower.

The flower is the striking beauty of a plant: the takeaway of comprehension.

But the ephemeral flower depends upon the leaves, which soak up light: the construals of data that render comprehension.

The leaves depend upon the stems: the relations of concepts that comprise comprehension.

The stems depend upon the roots: the assumptions that are the basis of comprehension.

The roots depend upon the soil: the paradigm by which comprehension arises.

Comprehension falters by pretext: appreciating the flower while ignoring the roots. This failure is the myopia of being stuck in the middle.

True comprehension is not of the flower, but of the entire plant and its habitat. Only in this soil does the garden of knowledge grow.

Commentary

Comprehension is an affirmative construal event. The mind decides that the snapshot it has taken is representative and a useful addition to its knowledge base.

When we look at a plant, we are first drawn to its flowers, which we consider the epitome of floral beauty. Flowers are indeed a beacon, albeit one that the plant intended for a select audience: pollinators.

Flowers are an ephemeral expression, a minor portion of a plant's life cycle. Comprehension is like a flower in taking what is most fascinating as apprehension. Especially for those with low awareness, comprehension focuses on the most striking, which is not necessarily the most accurate from a holistic perspective.

Comprehension depends upon one's reality paradigm. Believing in dualism or matterism invariably results in misunderstanding the proverbial plant.

Relying upon the wrong assumption about reality is the mental state of "being stuck in the middle": inability to get to the root of reality. Science turns into fiction with the blithe assumption of dualism or matterism, by confusing artifacts for energetic dynamics. American writer Hellen Keller observed, "The bulk of the world's knowledge is an imaginary construction."

A reality paradigm settles into the mind, growing roots that pervade perspective. Misconception is not easily removed, even if later found to be a worthless weed. This is especially true for those in iğnorance, who lack mental fluidity.

Knowledge is not an amalgamation of myopia. Genuine comprehension is holistic. A genuine philosophy is consistent at every scale of conceptualization.

English philosopher and economist Adam Smith wrote in the 18th century, "the theory that can absorb the greatest number of facts is the one that must rule all observation." That theory is *energyism*, the reality paradigm which underlies the philosophy in this book.

Once one has the correct paradigm, facts fit together into a verified coherence. Nature's display and dynamics are viewed as composing a wondrous whole.

Perspective

視点

Existence is a dance of perspectives.

Strength has a fractal perspective.

Problems are only solved by changing perspective. Others are only understood by changing perspective.

Empathy has twin forks. The soft fork resonates into compassion. From the sharp fork emanates power.

Fluidity in perspective reflects strength.

Commentary

Existence is a dance of perspectives.

A *reference frame* is a mental perspective. Underlying physics, chemistry, and biology is the idea that all entities operate as fields, with reference frames as bases to behave.

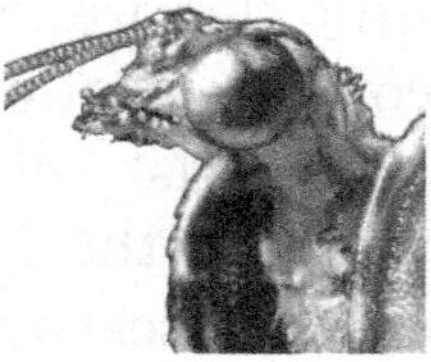

Mantis Fly
(photo courtesy of Ron Sterling)

Many arthropods have compound eyes. These eyes comprise many thousands of ommatidia: vision receptors. Each ommatidium has its own perspective. Via individual captures of light from ommatidia, an arthropod's mind composes a panoramic collage that lets it expansively see all around its body. Compound eyes exemplify a fractal perceptual process of unification via diversity.

Every field, whether quantum or molecule or cell or being, has its own perspective. The expression of Cöcö dynamically composes existence moment by moment using data from all these infinite perspectives. This incalculable miracle manufactures the mundane. American physicist Brian Swingle said of this dynamic, "Space is made of entanglement."

Strength has a fractal perspective.

The mind has a manifold of contributions. Not all mental inputs are intrinsic. But all fractally reflect strength of consciousness.

You may recall at some time hearing an inner voice, seeing a mental image, or having an intuition that did not seem to come from your own mind. Most people have had such moments. Every culture since prehistory, and most religions, have teachings about paranormal spirits making their presence felt.

Indian guru Nisargadatta Maharaj put a positive spin on spirits when he said, "You are never alone. There are powers and presences who serve you all the time most faithfully. You may or may not perceive them. Nevertheless, they are real and active."

At the other end of the presence spectrum are evil spirits. There are exorcism rituals among indigenous cultures worldwide and in almost all religions.

Plants and animals have microbiomes. A microbiome comprises the microbes – bacteria, fungi, protists, and viruses – that live within or on an organism.

Microbiomes are instrumental in the development and dobe of their hosts. Within you is a digestive microbiome that eats your food and feeds you the leftovers. Your appetite and food choices are outcomes of inspiration from your gut microbiota.

The collection of a person's presences are like a microbiome. A spiritbiome reflects mental strength.

The microbiome and spiritbiome are physical and spiritual counterparts. Holistically, strength is fractal in its quality. The microbiome and spiritbiome are part of that fractal. A soul is itself part of a larger fractal.

Many diseases result from a bad immune response. The immune system misbehaves, attacking the body rather than defending it.

The covid pandemic exemplified. For those with healthy immune responses, the virus itself caused little or no bodily distress. But for those whose homeodynamic health had been compromised, covid could be debilitating and long-lasting. Severe covid is an autoimmune disease set off by an otherwise mild cold virus.

Presences feed or feed off psychic energy. Benevolent ones help. Bad ones drain and sour a soul.

Presences make themselves felt in various ways: transient feelings, moods, thoughts, insights, strange dreams. Their mental polarity reflect one's own spirit.

Madness is a feeding frenzy of presences ravaging a soul. You may have heard deranged people having loud, disturbed conversations with their presences. Their nattermind is the leader of a psychically rancid band. These poor souls are being eaten alive from the inside out.

Strength acts as a spiritual immune system, with profound influence on health holistically.

Problems are only solved by changing perspective.

Overcoming an obstacle can only be done by changing mental method. Albert Einstein, who incorporated the idea of reference frames into physics with his relativity theories, said, "We cannot solve our problems with the same thinking we used when we created them."

Others are only understood by changing perspective.

Mentalizing is inferring the mental state of someone else. This empathic process of mind perception transpires by shifting psychological reference frame from the nominal perspective of oneself into that of another: how another person views a situation or dynamic.

Empathy has twin forks. The soft fork resonates into compassion. From the sharp fork emanates power.

Empathy is the essential psychological tool of social comity. There are 2 reference frames ("forks") of empathy.

The "soft fork" is emotive: mentalizing another's emotional state. This shifting of emotional perspective – emotively identifying with another – is the psychological basis of bonding between individuals. As a vital tool for infants, emotive empathy is inborn.

Our minds have an innate emotive response that evokes sorrow upon feeling a loss. Compassion is based upon this natural reflex.

The "sharp fork" of mentalizing is comprehending someone else's perspective via their desires. This knowledge grants the power to influence and manipulate others, especially the weak-minded (those lacking strength), who are easily swayed. Cognitive empathy is largely a learned skill.

Fluidity in perspective reflects strength.

"Fluidity in perspective" refers to the ease with which the mind moves in psychospace or changes mental reference frame, such as with empathy.

Those with less awareness have trouble changing their perspective. Being in thrall to nattermind makes for sticky mentation.

Those with higher awareness go with the flow, adapting their perspective to correspond with fresh information. An elevated consciousness level offers better faculty for introspection, including motive, and how and why certain mentation transpires. Another aspect of strength is being better attuned to intuition as a source of information, insight, and advice.

The Cage

籠

Consciousness is caged by the mind.
The bars of the cage are built of belief.
Consciousness seeks its freedom.
One may try experience as the key to the cage.
But doing is not key.
One may try reason as the key to the cage.
But logic is not key.
The key to the cage is will.

Commentary

Consciousness is caged by the mind.
The bars of the cage are built of belief.

Consciousness is the willful witness of the mind's productions. What the mind outputs depends upon biology and strength.

Purely biological perception is *coremind*. In making sense of sensation, coremind manages raw perception. Analytic by nature, coremind also solves problems.

Coremind works subconsciously: mentation which does not rise to overt awareness. Nothing passing through coremind is conscious. Coremind itself provokes no thought, no emotion.

Numerous adaptations crafted the human ape. This altricial animal is naturally social, but also willful and individualistic.

As a counterbalance to its independent streak, the hominin ape gained greater gullibility in its descent. Easy trust abets sociality: fostering group cohesion and somewhat lessening aggression, especially in uppity males.

Gullibility eases deceit. People are poor lie detectors. Humans are literally worse than dogs when it comes to trusting known liars.

To lend vividness to its renderings, the mind liberally mixes memory and imagination into its construals. This inspires belief in those of low awareness.

Belief is reification: the mental habit of treating abstractions as true. "Concepts that have proven useful easily achieve such an authority over us that we forget their earthly origins and accept them as unalterable givens," observed Albert Einstein.

The Collective are immersed in their beliefs. American psychologist William James noted, "Belief creates the actual fact." German philosopher Johann Wolfgang von Goethe saw, "Man is made by his belief. As he believes, so he is."

Wile is a nattermind forte. Nattermind compellingly touts its troublesome fabrications, invoking worry, anxiety, and other negativity.

Credulity gives nattermind an advantage in instilling its deceitful sway. Innate gullibility levels the battlefield between willmind and nattermind.

Nattermind fortifies beliefs by making them self-reinforcing. "The more we examine our beliefs and explain how they *might* be true, the more closed we become to challenging information," observed American psychologist David Myers.

Belief is an instrument of suffering and symptomatic of slavery to nattermind. Belief forges the chains that binds one to iğnorance.

Consciousness seeks its freedom.

Willmind is the mental expression of consciousness. Consciousness wants a pure view without nattermind corruption.

This intent is commonly called "spirituality." The desire for pure awareness arms willmind in its war on nattermind. When the will to freedom is steeled into resolve, spirituality is weaponized.

> One may try experience as the key to the cage. But doing is not key.

One may experience much and know little. Curiosity is the spark that lights learning, not doing per se. More saliently, doing does not necessarily affect quality of beĩng.

> One may try reason as the key to the cage. But logic is not key.

Reason is logic which pleases. *Philosophy* is a system of reason. A philosophy comprises a set of consistent definitions: consistent in the sense of not having contradictions within the system.

For a philosophy to be valid, the terms which delineate that philosophy must correspond with each other and holistically explain the system to which the philosophy pertains. Otherwise, a philosophy is either incomplete (lacking comprehensive explanation) or invalid (lacking consistency among terms in the philosophy).

Concepts are abstractions. While useful for skill or comprehension, concepts cannot elevate awareness.

> The key to the cage is will.

Exercising resolute will is the only way to escape the cage that monkey-mind builds to keep a consciousness iğnorant. The will is to quietude: to silence nattermind.

Now

今

All that exists is now. But glimpses of now are fleeting. The mind clouds now with the past and the future. Emotion fortifies the fog.

Commentary

> All that exists is now.

What is indisputable is that all of existence is always and only data about the present moment.

> But glimpses of now are fleeting. The mind clouds now with the past and the future.

Now is not what it seems. Our senses are not up to the task of rendering high fidelity. Vision is exemplary.

Your mind has you believe that your eyes are capturing a detailed video of the proceedings before you. The science of sight tells a different story.

From a particle perspective, light comprises photons. A photon is a packetized field. We see when our eyes catch photons that have bounced off objects.

Every time a photon encounters a fermion (matter quantum), its story changes. If an atom does not eat a photon whole as an energetic meal, it selectively snacks on specific photonic wavelengths. The photons that get away can tattle about a distant object that they met. By what light data was *not* absorbed, object luminance and color are conveyed.

The human eye can detect a single photon, though it takes 3 to sense a flash of light. A little light goes a long way in conveying enveloped photonic data.

Sight is reflective absorption and translation. That simple statement masks an astounding real-time transformation of intricate data into a visual map of mind-boggling complexity in its construction. "Our perception

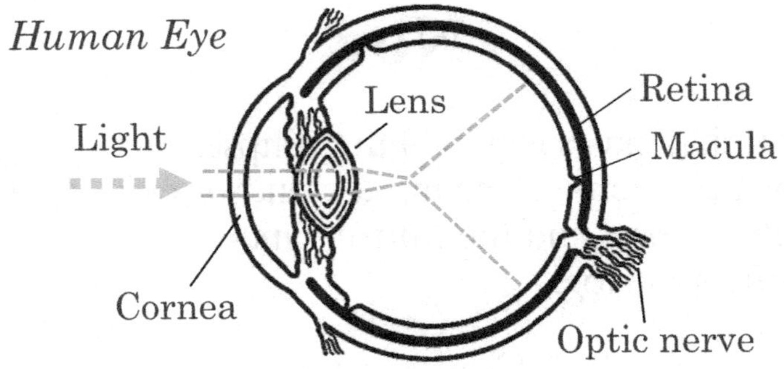

of the world seems so effortless that we take it for granted. But think of what is involved when you look at even the simplest visual scene. You are given 2 tiny, upside-down images in your eyeballs, yet what you see is a unified 3-dimensional world. This phenomenon is nothing short of a miracle," write Indian neurobiologists Chaipat Chunharas & Vilayanur Ramachandran.

"If we actually saw what our eyes take in, the world would be a chaotic place," observe American cognitive psychologists Michael Hout & Stephen Goldinger.

Every incoming spray of photons is bent by the eye's lens, and then strangely splayed on the retina, which is at the back of the eye. The retina comprises the rounded plane of light-sensitive tissue which absorbs the photonic impressers that enter your eye. The retina is the basis for rendering imagery. "Your retina is a terrible imaging device. Your only high-resolution vision is in the very center of your eye – about 0.1% of your entire visual field (the macula). You are *legally blind* to objects more than a finger width or 2 from the center of your vision," explain American neurobiologists Stephen Macknik & Susana Martinez-Conde.

Before being absorbed by the cells that transmit image data, photons must first pass through the nerve tissues that act as signal transmitters. The specific cells that absorb light are at the back of the retina. Whereas cone cells capture color, rods are only sensitive to luminance (relative brightness).

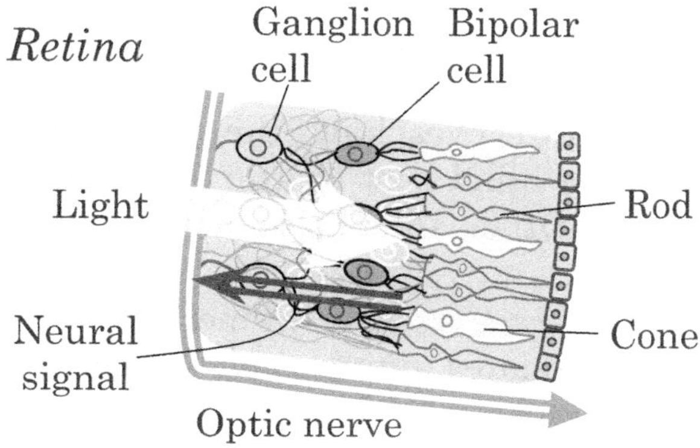

Given that a photon's story changes with every fermion it encounters – the very nature by which vision works – how light patterns that strike the cornea make their way to the back of the retina unmolested is inexplicable. While retinal tissue layers are largely transparent, cells that sit in front of the backplane rods and cones obstruct incoming photons, doubtlessly altering photonic field data.

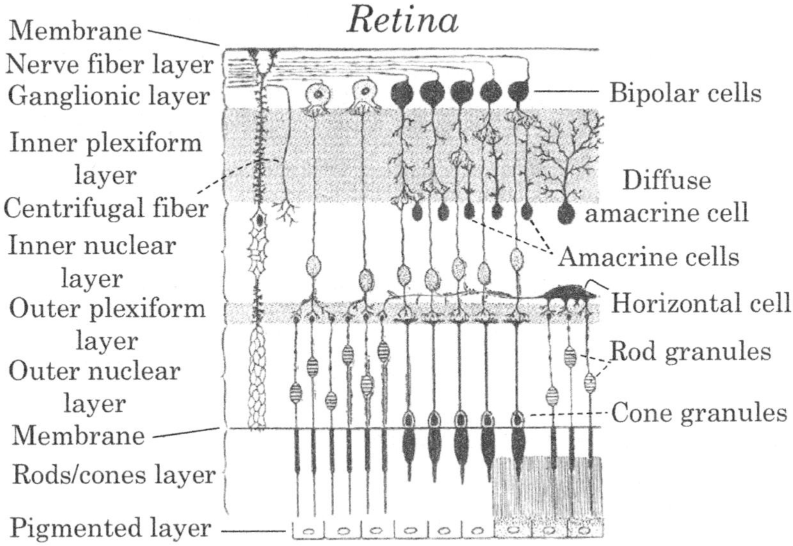

The physiological construction of our eyes has a "blind spot," where the optic nerve goes from the retina to the brain, leaving a hole in the fabric of sensory input. "Our perception in the blind spot is unreliable," notes German cognitive psychologist Benedikt Ehinger.

Only vertebrate eyes have the oddity of an optic nerve creating a blind spot by running in front of the photoreceptors. Other animals do not have this design defect.

So, how do we see so well? In mere fractions of seconds, the mind collates innumerable visual impressions and fabricates a colossal collage of stunning detail using memory and imagination. "It is quite clear that many aspects of our visual system are subconscious," notes American neurobiologist David Berson. "A large part is illusion. For many basic visual features, this 'filling in' is a general, and fundamental, perceptual mechanism," furthers Dutch psychologist Marte Otten.

This instantaneous impressionist painting has a compelling veracity. Experiments by Benedikt Ehinger and colleagues revealed that "generated information is sometimes treated as more reliable than sensory information from the outside world." Vision illustrates how effortlessly the mind's fabrications are believed.

A paucity of data from all our senses are massaged and augmented by the mind into compelling artifices that we naturally take as fact. Memory and imagination are lavishly employed to fill in gaps. The patterns heard in sound are a fabricated temporal collage.

It is hard indeed not to believe your own eyes. Therein lies the problem. Sight highlights that belief is a slippery slope. Perception completes the picture of deception.

Perception is generalization: seeing uniqueness and then categorizing novelties into mental schemas that inform how "the world" works. Perception fabricates concepts by subconsciously tailoring experience: pruning details by interest and level of awareness. These experienced abstractions are naturally taken as fact.

Once concepts of any sort are believed true, what bars the door to falling for fantasy wholesale? Nothing at all.

All knowledge comprises cerebral sandcastles of ideas. Such fabrications are accepted as Nature, with clouds of concepts providing the nourishing rain of "facts" that keeps knowledge growing. Amid this inner bustle, "glimpses of now" are indeed "fleeting."

Emotion fortifies the fog.

Emotion is monkey-mind's strongest stratagem. Take an intuitive feeling that may be informative. Amplify that feeling. Let the mind be overwhelmed with it, and even reverberate the body with it. Such is emotion.

Lost amid emotion is the bit of now which sparked it. Like a cerebral beast, whatever productive use a feeling may have had transforms into a rampaging monster with emotion. Emotion is a fog that blots out now.

Those in iğnorance are most in thrall to nattermind precisely because emotions are exhilarating. Performers in the arts are adored because they evoke emotion.

To one in iğnorance, entertainment is exciting. Calm is abandoned. Now is routed for thrill.

To a sage, entertainment is at most amusement. Calm remains.

Purpose

目的

The purpose of purpose is order.

The purpose of existence is its experience.

The purpose of living is its enjoyment.

The purpose of perception is fullness.

The purpose of knowledge is utility.

The purpose of memory is skill.

The purpose of imagination is purpose.

The purpose of noise is to instill the desire for silence.

Commentary

The purpose of purpose is order.

Purpose envelops intention: the will to a specific outcome. The antithesis of purpose is random.

Random is inherently unpredictable. Random is an attribution made when a pattern cannot be discerned.

Random and purpose are construals made from a willful perspective: from trying to comprehend and to create an outcome, respectively.

We may see an animal foraging and conclude that it is doing so randomly. Yet foraging patterns are optimized, either innately or with the addition of learning.

Complex mathematics are necessary to characterize foraging theory. Animals act out an intricate algorithm when they forage. Further, from the onset, an animal forages with a purpose.

Chemical reactions may appear to have some randomness. Yet every entity has behavioral rules it follows. This is true of field behavior from the quantum level up. The dance of electrons is specific to the music of the fields around them.

Nothing is random. Everything proceeds according to a rule set which is either intrinsic or purposeful. Such is the exhibition of cöherence.

Something considered random is an admission by a mind that it does not know what is going on. The intricacies of existence are often, by design, beyond ken to any mind.

The Cöcö-inspired process sees all, knows all. By this, existence happens in an orderly way.

The purpose of existence is its experience.

Cöcö is an ouroboros. Via localization, Cönsciousness witnesses the show put on by cöherence. The intent of Cönsciousness is to interactively experience what cöherence presents.

The purpose of living is its enjoyment.

Beings want to keep living for the sheer experience of it. This is ikigai.

Enjoyment gives living meaning. Meaning gives living purpose.

A consciousness expresses its interest via willmind. Its purpose is enjoyable experience. In reaping meaning, experience yields purpose, which renders purpose itself an ouroboros.

The purpose of perception is fullness.

Nattermind thwarts willmind by distraction which lessens awareness. Perception is diminished.

As consciousness is an interested witness, its nature is to take in as much as it can. Consciousness desires full, pure awareness.

The purpose of knowledge is utility.

The term *utility* means "of use." Applying will to dobe is *use*. Poor use is ineptness. Apt use is skill.

Competent doing requires knowledge. But so too the quality of beĩng. The more one knows of something, the richer the experience of it becomes. Whether doing or beĩng, the purpose of knowledge is utility.

The purpose of memory is skill.

Nattermind keeps the iğnorant in thrall by promoting memory for sentiment and thereby emotive attachment. Sentimentality is a corruption of memory.

Learning is the acquisition of knowledge. Knowledge becomes skill by recalling what has been learned and applying it successfully. Learning and knowledge are an ouroboros that yields skill.

The purpose of imagination is purpose.

Setting goals and solving problems requires imagining what is not to alter what is with the purpose of specific change.

Imagination creates purpose. When you set a goal, you establish purpose. When you solve a problem, you reaffirm intent to the goal that has been set.

A master of mispurpose, nattermind keeps the iğnorant in thrall by using imagination for purposeless fantasies, fostering reification.

The purpose of noise is to instill the desire for silence.

Nattermind manufactures mental noise in a multitude of expressions: distraction, negativity, stress, fear, worry, anxiety, depression, delusion. When suffering this noise becomes insufferable, it propels one to seek relief in quietude. The inyō of nattermind is that it inspires enlightenment.

The Collective

通常の人

The state and fate of the world reflect the Collective.

The Collective are possessed by their possessions.

Collective science is artifactual.

Collective authority is exploitative.

The Collective fight what they should accept and accept what they should fight.

Commentary

The state and fate of the world reflect the Collective.

The Collective are ordinary folk. The trait they all share is suffering the predations of nattermind.

The Collective dobe is expressed by culture, which includes language, expressed values, and political economy. In every ethnicity worldwide, culture expresses the Collective level of consciousness.

American singer James Brown soulfully sang, "This is a man's world." That is because women have been subjected in societies worldwide since the advent of agriculture 13,000 years ago, when physical strength and stamina surpassed savvy as the survival edge. Men plowing fields plowed women under.

Edicts for misogyny are vividly portrayed in religious texts which became the immoral mandate of societies worldwide. Christianity and Islam are notably steeped in this tradition. Since history began, every woman has felt the sting of subjugation from men.

The effect on humanity of man's social domination has been incalculable. The most poignant point on this score is that the incessant conflicts which have driven the currents of history would not have been if women had been running humanity's affairs.

Cooperation enabled all the achievements that embodied the ethical aspects of civilization. With men in charge, those victories have been hard-won. This remains obvious in the dynamics of political affairs that comprise world news. The comity which comes naturally to women is a compromise to men, whose sense of territoriality and possessive appetites have made civilization a savage endeavor since it began.

The modern world is largely ruled via democracies, where the public elects how its society is governed. It is apparent that democracy translates into a political sinking to the lowest common denominator. This is most apparent in the USA and UK, which have elected leaders for decades that led only in presiding over their nations' decay.

In 1648, Swedish statesman Axel Oxenstierna observed, "Behold with how little wisdom the world is governed." Keen observers of current events appreciate the historical constant in that denouement. Civilization has been a continuum. Its modus operandi of exploitation and social dominance hierarchies has never altered.

Fate is cause and effect writ large. The idea of fate is antediluvian. The word exists in every language. Yet, via propaganda by authorities, modern cultures outmoded the concept.

The touted substitute for fate is a fantasy of fluidity via faith in technology and economics to quickly effect solutions. This fantastic optimism belies history.

Technology has been a great irony. Technologies brought fantastic conveniences which changed the way we live and irrevocably shaped cultures. The pollution from those same technologies has caused atrocious misery and been so dramatic as to sour the environmental dynamism of Earth itself.

"What might now shower immeasurable material blessings upon mankind may bring about its total destruction," presciently foresaw British war commander Winston Churchill in the middle of the last century. That destiny is now being enacted through extreme weather events which have become the norm worldwide. That the end times are upon us is obvious for those with sufficient detachment to see the current mass extinction event for what it is. *The Story of Humanity: Ecology & Consequence* by Ishi Nobu relates humanity's history and fate.

The Collective are possessed by their possessions.

The human mind innately conceptualizes Nature as a diversity of objects. All human languages are object-oriented. Nouns outnumber verbs in all vocabularies. Infants learn new nouns easier and more rapidly than new verbs. Inborn objectification is integral to building the primal illusion of physicality.

Beyond object orientation is attachment to objects. People attribute ownership to object representations from early childhood.

Like many animals, humans are territorial. Materialism is the votive perspective of territoriality writ small.

For those in the Collective, possessions are a major contributor to and reflector of self-identity. American psychologist William James wrote, "A man's self is the sum total of all that he can call his."

American marketing professor Aric Rindfleisch got to the root of the problem: "At its core, materialism is a value-based response to insecurity in one's life." Rapt over physicality, the Collective try to fill their spiritual vacuum with possessions.

Collective science is artifactual.

In antiquity, *science* was a synonym for knowledge. *Science* is now used for the study of Nature.

The nature of scientific inquiry is induction. Physical artifacts and their movements are observed. Instance observations are conceptually generalized into characteristics and behaviors for an assigned system. The invariable result is a reification which Collective scientists fail to recognize as such.

Studied specimens are taken to be part of an objective reality. By this matterist assumption, Collective science is stuck in the middle.

The indisputable finding of modern physics – that matter is energy and energy is purely conceptual – is disregarded, even by physicists. In assuming artifacts as authentic, modern science is fiction.

Collective authority is exploitative.

Dominance hierarchies are seen in many animals. Authority is commonly exercised as subjugation, with subordinates serving the wants of a dominant.

Political states maintain order by force, which is implicit only until push comes to shove. This observation has been a political constant since antiquity.

Capitalism is an extension of the biological bent to dominance. The English word *profit* derived from the Latin word *profectus*, meaning "to do good." From the 15th century, when capitalism was emergent, *profit* was perverted to mean "the advantage gained by owning capital."

Plutocrats vitiated numerous words to warp culture to their worldview. Derived from Latin, the 14th-century English word *equity* meant fairness. Capitalists transformed *equity* into a property claim in the late 19th century.

The nature of plutocracy is that dominants with resources command others by providing the means for subordinates' survival. Worker *employment* is a statement of subjugation. The inequities in societies and dire condition of Earth's natural environment owe to exploitative capitalism being the engine of economic activity.

> The Collective fight what they should accept and accept what they should fight.

The vast majority in modern societies struggle to survive, as plutocracy handicaps the working class. Yet, even in democracies, where people could elect moral equity, they instead vote for injustice. A generous interpretation is that they do so because plutocracy is deceptively sold as offering freedom, whereas socialism implies an unseemly societal orderliness. A more telling construal is that many do not know what is good for them.

Those with compassion hanker for justice and societal equity. They protest the evils of plutocracy.

Then there is the dire condition of Nature on Earth, which has suffered accelerating devastation since industrialization in the mid-19th century. Environmentalists lament the rape of Nature, as they have for millennia.

Alas, these crusaders for a better world are fighting for a lost cause, as history amply shows. The force of plutocracy runs deep and is not going to be remedied. Protest is mostly martyring one's own time and energy.

The war to be won is not political. It is spiritual.

One's purview is one's own mind. In there, nattermind is the nemesis. Noise and negativity are dispiriting, sapping a soul of psychic energy by drowning out bliss. Such is living under the regime of monkey-mind.

The Collective accept suffering under nattermind as expectable because they consider such abuse "normal." They pass on their inner abuse to others.

The Collective do not recognize that conscious thought robs awareness. By this theft, nattermind substitutes a poverty of abstractions for the richness that comes with beĩng in quietude.

Awareness and discipline are the essential tools for every endeavor. Living well is a skill.

Vanquishing nattermind is best done with a dual attack. Both vectors are nonintuitive, as they go against biological programming. It is because of this that elevating awareness is difficult.

1st, stop thinking. Nattermind is thwarted by discarding thought and relying on subconscious mentation (menition). This mental streamlining – living in meditation – permits full awareness to the present.

2nd, believe nothing. Innate inclination to gullibility serves ill. Practice skepticism.

There are no truths in abstraction. Skepticism acts as a solvent on falsities.

The residue of nattermind is scrubbed by discarding beliefs and assumptions. Stop being stuck in the middle. Get to the root.

The Collective do not elevate their awareness because they do not practice a disciplined regime of skepticism. They assume. They believe.

The mind is a malleable device. There is a world of difference between using a tool versus imparting mystical meaning to that tool by believing in it and thereby granting it power. You either use your mind as a tool or you are a tool of your mind.

All you control is your own attention. Every moment you either choose now or indulge monkey-mind by reveling in sentimental memories or imaginative fantasies. This is the constant battle of willmind versus nattermind. This is the fight that defines the quality of your life.

Values

価値

The highest value is like water.
The value in awareness is richness.
The value in a mind is clarity.
The value in action is beneficence.
The value in relations is benevolence.
The value in words is sincerity.
The value in effort is timeliness.
The value in work is competence.
The value in management is expectation.
The value in leadership is order.
The value in economy is economy.
The value in authority is empowerment.

Commentary

The highest value is like water.

Water shapes the world. Water is essential to Life.
In its inherent fluidity, water does not contend.
Conceptualization is essential to living.
Conceptualization shapes perspective.
A *value* is an appreciated conceptualization.

Concepts worthy of esteem are like water. They do not contend. They fluidly fit the situation to bring insight.

Concepts are most valuable when they coalesce into a philosophy. Like water in the physical world, concepts shape the inner world.

The following passages are values attained with strength.

The value in awareness is richness.

Like water, purity of awareness affords an accurate view of what is going on. Polluted by nattermind, awareness is foggy and myopic in iğnorance. Awareness blossoms in the mental transformation of enlightenment. To achieve rich awareness, nattermind must be vanquished.

The value in a mind is clarity.

Confusion is common in iğnorance. Indulging nattermind grows the weeds of mental illness, crowding out the yield of lucidity.

Mental clarity is achieved by discarding thought and belief, and refusing thrills, including the psychological charge that occurs in indulging negative emotion.

The value in action is beneficence.

Goodness often generates positive feedback in ways that cannot be anticipated. To effortlessly achieve beneficence, one must be imbued with bliss. This happens naturally upon ascending to higher awareness.

The value in relations is benevolence.

Benevolence results from empathic mentalizing. To achieve benevolence, one must comprehend human nature and in doing so feel compassion for those who suffer in iğnorance. The perceptivity and bliss garnered via higher awareness facilitates benevolence.

The value in words is sincerity.

Communication offers the opportunity of social bonding. To achieve sincerity, one must be content, not wanting from others.

The value in effort is timeliness.

Efficient effort is most rewarding. To achieve timeliness, one must hold in mind both skill from the past and the goal of the future while dobe resides in now. Higher awareness improves focus.

The value in work is competence.

To achieve competence, one must know the ideal result. As such, imagination is essential to competence. Higher awareness affords clarity in distinguishing the essential from the inessential.

The value in management is expectation.

A team endeavor can only succeed if all members perform their role. Each member must know what is expected of them. A manager achieves expectation by knowing all roles and communicating them to each member so that the project completes as planned.

The value in leadership is order.

Leading others is like cooking a small fish. Too much stirring will make a delicate fish fall apart. Too little stirring will leave the cooking uneven. To achieve order, one must know the entire structure and dynamic.

The value in economy is economy.

This passage twines the value of doing in double meaning. An *economy* is a system of producing and distributing for survival of people in a society. Chronic waste is counterproductive and signals that an economy is not sustainable.

To achieve frugality (economy), an economic system must be regulated. Only by this could all in a society prosper and Nature be preserved.

Enlightened leadership and management are required for an economical economy. Such a political economy, obviously, is not the way of the Collective world.

The value in authority is empowerment.

Order is best obtained when society's members see order as befitting their own interest. To rule by force leads to resentment, paving the road to revolution.

A prosperous society requires more than order.

Proper authority begins with apt leadership. Apt authority provides the means for society's members to make the most of their lives. To enliven empowerment, authority must appreciate the many ways that enrich a society.

As with the economy, proper political authority has been a rarity in the world of the Collective. Discrimination on a tribal basis has long been the norm.

The War Within

戦争以内

A war is waged for a soul's destiny.
The war cannot be won through achievement.
The war cannot be won through detachment.
The war cannot be won through esteem.
The war cannot be won through thought.
The war cannot be won through faith.
The war cannot be won through knowledge.
The war can only be won by silence.

Commentary

A war is waged for a soul's destiny.

A transcendental soul incarnates as a willful witness to the proceedings of living. The mental expression of consciousness is *willmind*.

To make living a challenge, cöherence contrives a nemesis: *nattermind*. Banishing this bane is what willmind must do to shape a better destiny, both while alive and for incarnations to come.

The war willmind wages is for higher awareness. Enlightenment is the first stage of clarity, when abiding bliss presents itself as a vehicle to savor living. Higher levels of consciousness further enrich the experience of living.

The war is between willmind and nattermind. Nattermind wants consciousness mired in iğnorance, wallowing in suffering.

The war cannot be won through achievement.

Enlightenment is not of doing, but of a refined beĩng.

The war cannot be won through detachment.

Attachment is emotively clinging to conceptualizations for psychological comfort. Attachment is a salve for insecurity. The compulsion to attachment is strong in iğnorance. Enlightenment cannot be attained by pretending nothing matters.

The war cannot be won through esteem.

One's esteemed values are irrelevant to awareness level. There is no morality to enlightenment.

The war cannot be won through thought.

Now is all that exists. Thought is the robbery of the present moment from awareness.

Thought is useless. Vasistha, the ancient Indian guru who wrote seminal Vedic texts, observed, "Though appearing to be intelligent, thought is unable to comprehend anything really."

The war cannot be won through faith.

Faith is attachment to certain beliefs. *Belief* is the esteem of concepts. Belief supplies no bullets in the war against iğnorance.

The war cannot be won through knowledge.

Enlightenment is not of the mind or of concepts but of awareness. It does not matter what is known.

The war can only be won by silence.

Enlightenment can only be had by subduing nattermind. The will to quietude is the winning strategy.

Ways

道

When spiritual seekers hear of Cöcö
they diligently try to discover its way.

When ordinary people hear of Cöcö
they appear both aware and unaware of it. They long for spirituality but are captivated by the material.

When dullards hear of Cöcö they think of it as God. They believe religion is spirituality.

Those realized live as Cöcö incarnate. Their way is like water, rising and receding as opportunity flows. They cling to nothing.

Commentary

We have an innate yearning to enjoy life to its fullest. This abiding desire is the wellspring of spirituality.

People respond to mention of reality and elevating awareness according to their spiritual aspiration.

Those already on their spiritual journey seek to further elevate their consciousness. They welcome more information.

By contrast, most in the Collective respond to news about mental health in various ways. Some want a deeper experience in living ("long for spirituality") but remain under sway of nattermind ("captivated by the material").

Those of dim awareness ("dullards"), in thrall to their monkey-mind, believe in God. They do so from fear, wanting salvation from their inner torments. Such grandiose fantasy mistakes object for process. Whereas Cöcö signifies the necessary processes of reality that manifest in Nature, God is the idea of a supreme overlord.

Everything known about Nature, from physics to biology to psychology, validates energyism: the reality-to-existence mechanics of Cöcö as explained herein. By stark contrast, there is no evidence for an incorporeal supreme being, especially one specifically concerned about human morality.

Gullible simpletons lap up religious dogmas and are fervent adherents to ideologies: true believers in many mental realms.

Curiosity is the impetus to further comprehension. "Curiosity is the wick in the candle of learning," wrote American writer William Arthur Ward.

Curiosity arises from surprise, a feeling of nescience, or confusion. Surprise at something that belies current comprehension provokes curiosity as remedy. Curiosity motivates closing the gap between what one knows and what one would like to know. "Uncertainty – when you think you know something and discover you don't – leads to the most curiosity and learning," observed American psychologist Celeste Kidd.

In providing facile answers, belief throttles curiosity. Belief is lazy mindedness.

Sages have a purified consciousness that is harmonized with Cöcö. They live fluidly ("like water"), dealing with whatever situation they are in, while prudently taking advantage of opportunities which arise. Objects are conveniences, not treasured possessions.

Sages spurn belief. By this they stay open to revising their science based on fresh information.

Will & Strength

意志及強度

Will is the kindle of strength. The fire sparked is light without heat.

The power of will is in immersion.

Strength is like water without wind. Stillness creates waves.

Strength is expansion via subtraction.

Strength results in full doing from pure beĩng.

Strength results in fluidity, but not by will.

Strength results in detachment, but not by will.

Strength results in insight, but not by will.

Strength results in bliss, but not by will.

Commentary

Will is the kindle of strength. The fire sparked is light without heat.

Strength is perceptivity and lucidity. Will is propellant to strength. The "fire" of strength is "light without heat": mental clarity without distraction.

The power of will is in immersion.

The "immersion" of will is discipline. Habitual will to quietude develops strength, which blossoms into power.

Strength is like water without wind. Stillness creates waves.

Quietude results in mental clarity. The "stillness" of inner silence permits the most power ("waves").

The physics analogy of psychological strength is *potential energy*, which may be compared to the *kinetic energy* of psychological power.

Strength is expansion via subtraction.

Perceptivity ("strength") affords greater awareness ("expansion"). This is done by eliminating the noise of thought ("subtraction"). Strength comes by relying upon the magic of subconscious mentation.

The passages in the rest of the chapter explore what results in attaining strength. These benefits cannot be summoned ("not by will"). Instead, these process characteristics naturally occur from acquiring strength. In context, "naturally" means a consciousness gaining greater conductivity with Cönsciousness, which is the root of ultimate power.

Strength results in full doing from pure beĩng.

Strength is "pure beĩng," which is quietude when acting as an unengaged witnessing consciousness. Power, which results from strength, allows "full doing."

Strength results in fluidity, but not by will.

In context, "fluidity" refers to productive movement in psychospace. Psychospace can, of course, be traversed by will. But once one has strength, the mind moves efficaciously through psychospace effortlessly.

Strength results in detachment, but not by will.

Attachment, which nattermind inspires, weakens with strength. One cannot easily stop caring about what one is attached to.

Strength results in insight, but not by will.

Insight is valued intuition. There is no way to summon insight. But insight comes easily when sought with strength. This occurs as consciousness becomes a clearer conduit with Cönsciousness.

Strength results in bliss, but not by will.

Bliss comes with quietude: a prime benefit of being harmonized with the unified field of Cönsciousness.

Transformation

变換

Vanquishing nattermind begins a transformation toward the divine.

Quietude begets bliss.

Bliss is the light of enlightenment.

The calling card of cöherence consciousness is cöherence revealed.

Realization is the epitome of divinity.

One in iğnorance cannot imagine realization. There are no words to describe experience beyond words.

Commentary

Vanquishing nattermind begins a transformation toward the divine.

A field of consciousness is a fractal process portion of the universal field of Cönsciousness ("the divine").

There are 3 progressive stages of higher awareness: enlightenment, cöherence consciousness, and realization.

The inner transformation to realization begins with quietude: breaking with the past of believing what comes to mind to refusing what comes to mind.

Elevating awareness level enlivens clarity as the fog of nattermind dissipates. Abiding clarity forges a harmonization of consciousness with Cönsciousness. The most noticeable symptom of this is the strong waves of bliss felt with the advent of enlightenment.

The gravitational interaction of the spirit transforms with elevating awareness. Psychologically, *spirit* is the vitality which willmind identifies as the dobe aspect of *self*. The axis of dobe, and hence sense of self, shifts as awareness ascends.

In ignorance, nattermind convinces willmind to be self-centered. The dobe orbit is selfish. Mentalizing is often used as a tool for achieving one's wants.

Elevating consciousness affords greater environmental awareness. Socially, patterns of dobe become increasingly construed as a web of relations. As a result, mentalizing naturally becomes more compassionate. This transformation is facilitated as attachment dissolves. This includes self-centeredness, which is attachment to one's sense of self.

With the blessing of bliss, a spirit's aura brightens as awareness rises. Not holding onto negativity yields a lightness of being that others perceive. Positivity heightens social power.

> Quietude begets bliss.
> Bliss is the light of enlightenment.

The overwhelming lightness in enlightenment is the bliss which is felt from Cönsciousness.

Bliss begets better health, both mentally and physically. Stress no longer plagues mind or body.

> The calling card of cöherence consciousness is cöherence revealed.

A good contextual assessment is known as having "the big picture." The idiom, "the devil is in the details" highlights what English musician Pete Townshend pointed out in a song lyric: "The simple things you see are all complicated." You get the big picture by inductively collating details into a contextual whole.

The unity of diversity in Nature becomes a visceral experience in cöherence consciousness. One is literally conscious of cöherence. Perceiving intricate patterns is a powerful benefit that occurs at this level of awareness. Dealing with the devil in the details and sussing the big picture are easier as awareness ascends.

Realization is the epitome of divinity.

Realization harmonizes consciousness with Cönsciousness. But being an animal is a coarse experience. In realization, one is as divine as animal incarnation affords – which, frankly, is not so sublime.

Biology is the ballast of Life that may be transcended only momentarily. But those moments are divine.

One in iğnorance cannot imagine realization. There are no words to describe experience beyond words.

Those in iğnorance may fantasize enlightenment as overflowing with positivity. There is more to enlightenment than bliss.

Clarity firms practicality. Judicious skepticism is the essential tool for practicality. Bliss merely helps staying emotively grounded.

In vanquishing nattermind, there is an abrupt transformation from iğnorance to enlightenment. By comparison, ascending from enlightenment to realization is a more gradual inner development.

There is a flush of bliss that occurs in the blooming of enlightenment. This wave helps cleanse the spirit.

Bliss is a psychic bandage. The struggle of living does not end. But, in higher consciousness, bliss softens the blows.

Ascending to realization is sobering. Bliss on tap cushions the cold, hard comprehension of existence as a transactional exercise: an experience ridden with deception that spells suffering for so many (the Collective) – by design!

Sages pity the iğnorant because those poor souls are unaware and suffer for their sad state. *Katha Upanishad*, an ancient Vedic text, summarized the Collective: "Fools dwelling in iğnorance, yet imagining themselves wise and learned, go round and round in crooked ways, like the blind led by the blind."

For those in higher consciousness, sociality is smoother because they rely upon grace and silence to get by among the Collective. Like mischievous crows, some sages may playfully try to inspire skepticism which may jostle someone from being "stuck in the middle."

A sage knows that love is just adoration of a concept. Having no attachments, a sage no longer feels the passion that those in iğnorance relish.

The foregoing description may make realization seem undesirable. That is only because the richness of realization is as indescribable as it is unimaginable to those in iğnorance. Inner transformation with elevating awareness is profound indeed.

Thanks to deep-seated bliss, enjoyment is generalized in realization. The game of life is enjoyed for what it is.

Sagacity

聡明

A sage behaves as a child with wisdom.
A sage influences without authority.
A sage plans yet behaves spontaneously.
A sage has skill without esteeming knowledge.
A sage feels contentment underlying discontentment.
A sage knows that the world is an illusion.
A sage believes nothing and experiences everything.
A sage transcends inyō.

Commentary

Realization, also called *unity consciousness*, is the highest level of awareness. At that stage, individual consciousness is as connected with Cönsciousness as biology affords.

A sage behaves as a child with wisdom.

A sage lives honestly and simply, enjoying physical pleasures but not attached to them.

The intricacies and patterns of cöherence show themselves with unparalleled clarity, inspiring awe.

In being harmonious with Cöcö, intuitions and insights are mundane. Thanks to perceptivity and mental fluidity, sages are naturally wise, even those who are academically unlearned.

A sage influences without authority.

A sage's aura of calm radiates and can be felt by others, even by those with dim awareness. The words and acts of one in realization impress others as exemplary.

A sage plans yet behaves spontaneously.

Goal-oriented preparation is a necessary aspect of living. Yet only by behaving spontaneously can intuition apply to the uniqueness that is every situation. Such is what a sage naturally does: retaining flexibility while working toward a goal.

A sage has skill without esteeming knowledge.

Skill is the application of knowledge. *Knowledge* is a conceptual classification.

A sage knows that abstraction is a shadow of truth, that the only value of knowledge is for its convenience in skill.

A sage feels contentment underlying discontentment.

Desire is an expression of discontentment: of wanting. As a sage is ballasted by bliss, contentment is still felt when discontent arises. This renders desires selective rather than compelling.

A sage knows that the world is an illusion.

Physics and biology have provided abundant evidence of *energyism*: that the mind creates the mirage of matter. Nature is deceptive by design. This scientific validation of Cöcö confirms what sages experience and gurus have taught since antiquity.

A sage experiences living as an extended dream: engaged but detached. The designs that comprise Nature are a wondrous perceptual buffet. The artifice of mentation is seen as a grand ruse.

A sage believes nothing and experiences everything.

A sage's mind is cleansed of belief. Though convenient for skill, abstractions are known to be inauthentic.

This purity allows a sage to experience every moment as unique, without the bias of trying to classify and thereby abstract away the delicacies that lace the beauty of Nature. The mind's concoctions are a confection.

A sage transcends inyō.

Dobe is the doing and beĩng that comprise living.

Doing arises from desire. Beĩng is pure witnessing: something which those in iğnorance have but brief glimpses of.

Desire compels those in iğnorance, who often tussle with purpose, meaning, and significance. Their desires sustain an inyō tension. This struggle occurs because nattermind clouds clarity. Many fear death because of this: not knowing what living is for and feeling attached to their perceived physicality.

Desire is transparent to a sage. Doing is merely a means to some convenient or aesthetic result. There is no "larger" meaning or purpose.

A sage lives in pure beĩng, which is imbued with bliss. Though pleasant, bliss is understood by a sage to be a ploy of Cönsciousness for ikigai.

A sage lives transcendentally, taking life as it comes without concern. A sage knows that incarnation is an endless wheel. Reaching the pinnacle of dobe paradoxically reveals that nothing really matters. For a sage, the inyō of dobe is resolved.

Glossary

A

absolute: pure; without limits; nonrelative. Contrast *relative.*

abstraction: a symbolic representation. Compare *concept.*

achieve: to successfully do. Compare *attain.*

act (verb): do; initiate behavior. Contrast *react.*

actual: existing.

actuality: experienced through the senses. Compare *reality.*

actualize: to make actual; to represent realistically; to give substance or performance capability to; to manifest in the broadest sense.

adaptation (evolutionary biology): the teleological (goal-oriented) process of adjusting to ecological circumstance. Adaptation aims at enhancing survival and reproduction while keeping living a challenge.

aka: also known as.

alacrity: quickness.

alchemy: the study of matter transmutation, which scientifically evolved into chemistry.

alive: a consciousness witnessing dobe. See *dobe.*

amalgamate: to unite or combine into a whole.

ambient: a surrounding or encompassing environment.

amusement: pleasant entertainment.

animate: alive. Contrast *inanimate.*

animism: the doctrine that a vital energetic force is inherent in all of Nature. Compare *vitalism.*

antediluvian: exceedingly old; (now conventionally considered) outdated.

antithetical: the exact opposite.

apperception: (awareness of) intuitive comprehension.

apprehension: an act of intuitive understanding.

archaea (singular: *archaeon*): a taxonomic domain of life alongside bacteria and viruses; the group of prokaryotes from which eukaryotes arose, by union with a mutualist bacterium. Archaea account for 20% of Earth's biomass. Archaea are extremely robust and versatile, with both extremophiles and ubiquity in their favor. Marine archaea are plentiful. The archaea in plankton make them among the most abundant organisms on the planet. Archaea play roles in the carbon cycle and nitrogen cycle. Typically gregarious, archaea are commonly mutualists or commensals. No archaeal pathogens or parasites are known.

artful: cunning; crafty; deceitful. *Artful* is one of those words that has an evolution of opposite meanings. From the 14th century, *artful* meant skillful (full of art). As used in this book, *artful* acquired its negative connotation by the late 16th century. The idiom "artful dodger" became popular in the 1560s. Contrast *artless.*

artifact: an object; a piece of tangible (material) evidence.

artifice: a clever or shrewd trick.

artless: sincere; simple; genuine. Contrast *artful.*

arthropod: an invertebrate with an exoskeleton, a segmented body, and jointed appendages. Arachnids, crustaceans, and insects are arthropods. Before the current mass extinction event accelerated in the late 19th century, there were over 6 million distinct arthropods, comprising over 75% of animal species. Their collective biomass far outweighed that of vertebrates.

artifact: a tangible creative work.

artifice: a clever ruse, trick, or contrivance.

assumption: supposing or taking for granted.

atom: the smallest unit of a chemical element. Atoms are comprised of subatomic quanta.

attachment (psychology): emotively clinging to conceptualizations for their construed psychological comfort: a compulsion of felt insecurity.

attain (psychology): to come to a state of beĩng. Compare *achieve.*

attention: willed mental awareness.

aura: the subtle, perceptible field emanation of an object or being.

authority: power over others.

autoimmunity: a bodily disease arising from an immune response that attacks the body rather than defending it.

autonomic: relating to subconscious intent.

awareness: being conscious. See *consciousness*.

Axial Age (8th–3rd century BCE): German philosopher Karl Jaspers' proposal that there was a pivotal age in world history regarding the development of philosophy and religion.

axiom: an assumption.

B

bacteria (singular: *bacterium*): a taxonomic domain of single-celled prokaryotes, abundant in most ecosystems. All animals rely upon mutualist bacteria for their digestion and to assist with other bodily functions. Bacteria play vital roles in various facets of the biosphere.

banish: to dismiss, exclude, evict.

baseline: a line or level that is nominal, minimal, or typical.

beauty: qualities which excite pleasure.

beget: generate; produce.

behave: to act or react in a certain way.

behavior: an attribution of action or inaction to a living entity.

being (noun): a living life form, whether a molecule (e.g., virus, protein), cell, or organism. Compare *beĩng*.

being (verb): to exist. Compare *beĩng*.

beĩng (noun which is really an intransitive verb): the intrinsic awareness of being alive; awareness in the living mode of *not* doing. The tilde ĩ disambiguates beĩng from other uses of being. See *doing*.

belief: a habit of the mind to treat ideas as true; confidence in abstractions as real.

beneficence: active goodness.

benevolence: good that benefits others.

bereft: lacking something needed.

bias: a subconscious preference.

biofilm: a colony of prokaryotes encased in a stabilizing polymer matrix; commonly known as *slime*.

black hole: (literally) a hole in the fabric of existence. Black holes have tremendous gravitational pull. Each galaxy rotates around a massive black hole. Cosmologists cannot explain how the cosmos evolved to be galaxies orbiting black holes.

bliss: the feeling of joyful contentment which emanates from connection with Cönsciousness. Bliss is symptomatic of enlightenment. Compare *happiness*.

blithe: carefree; unconcerned; casual or indifferent.

boson: a localized field which acts on fermions. See *cöforce*.

C

canticle: a poem, nonmetrical hymn, or song intended as inspirational.

capitalism: a market-based economic system based upon private ownership of resources and their exploitation for exclusive profit. Compare *socialism*.

category: a grouping by some perceived characteristic(s).

causality (aka (noun) *cause and effect*, (adjective) *cause-and-effect*): the construal that one phenomenon provokes a succeeding phenomenon. Contrast *correlation*.

causation: an imagined linkage of cause and effect.

cause (verb) (physics): to effect; to bring about.

cell (biology): the basic physical unit of organisms. Compare *virus*.

challenge: a provocative invitation to enter a struggle of skill.

character: the perceived distinguishing qualities or attributes (of something).

charge (electric): the force of electromagnetism per unit of time, measured in *coulombs*. Electrochemical charge is measured in *faraday*.

chemistry: the study of matter, especially chemical reactions.

clade (biological classification): a group of biological taxa, such as species, which includes all descendants of a common ancestor.

clarity (psychology): precision in perception, lucidity, perspicuity.

coarse (psychology, sociology): vulgar, crude, lacking refinement or sensitivity.

Cöcö: the universal, unified field of Cönsciousness/cöherence which localizes into individualized consciousnesses, minds, and the perception and production of Nature.

cöforce: a local field of effect without manifestation.

cognition: willful mentation to gain knowledge.

cognitive (adjective): related to mentation.

coherence (psychology): the perception of having order. Compare *cöherence*.

cöherence: the unified field which localizes and quantizes to produce minds and Nature.

cöherence consciousness: the 2nd level of higher consciousness, where the unity of Nature is sensed. Compare *enlightenment* (the 1st level), *realization* (the 3rd level).

Collective: those who live in iĝnorance, which is most people.

combustion (aka *burning*): a high-temperature exothermic redox chemical reaction between a fuel (the reductant) and an oxidant, usually oxygen. Combustion produces smoke, which is an oxidized, gaseous mixture. Combustion does not necessarily result in fire. A flame is only visible when substances undergoing combustion vaporize. The heat from a flame may provide enough energy to render combustion self-sustaining.

comity: mutual social consideration, including courtesy and respect.

compassion: a feeling of sympathy in witnessing misfortune, often accompanied by an emotional impulse toward helpful behaviors. See *empathy*.

competence: the state of being capable or sufficient for a task.

complement: something which completes or balances a system.

complementary: serving to complete or conceptually fill out.

compound eye: a type of eye comprising thousands of individual light receptors (ommatidia) from which the mind constructs imagery.

comprehension: contented understanding.

conation: volitional (willed) mentation.

concentration (psychology): pure focal awareness.

concept: a symbolic construct; a construed abstraction or idea.

conflation: blending or (con)fusing concepts.

confusion: lack of clarity, bewilderment, disorientation.

congruence, congruity: the quality of coinciding; in a state of agreement or correspondence; in harmony.

conscious: in awareness; aware of. Compare *subconscious*.

consciousness: a localized faculty for awareness. See *Cönsciousness*.

Cönsciousness: the unified field of consciousness which localizes into individualized consciousnesses.

consilience: concurrence of theories from distinct disciplines of knowledge into a set of inductive principles.

construal: (an) interpretation.

construct (noun; psychology): a conceptual system.

construe: interpret; comprehend meaning or intent.

consume, consumption: to allocate attention to.

content: satisfied; not wanting.

coremind: fundamental perception. Compare *nattermind*, *willmind*.

correlate: congruence in a relationship.

correlation: the fact that multiple phenomena coincide. Contrast *causality*.

cosmogony: the origin or coming into being of the astrophysical cosmos.

cosmology: the study of the astrophysical cosmos.

cosmos (astrophysics): the (seemingly physical) universe.

cosmos (philosophy): the totality of existence.

counterfactual: not existing.

counterfactual cognition: imagination.

covid: an autoimmune disease via CoV2 viral infection. CoV2 is a coronavirus.

credibility: the quality of being believable.

curiosity: a desire to know more (about something).

cytology: the study of biological cells.

D

data: symbols which may become meaningful via perception. Compare *information.*

daydream: idle imagining. See *dream.*

deduction (logic): the method of inferring a conclusion about particulars from general principles. Contrast *induction.*

desire: (a) want. Compare *need.*

device: a designed tool; a trick or scheme.

dipole (*dipolar*): having 2 poles. Compare *monopole.*

discipline (verb): to bring or have under control.

distal: distant. Compare *proximal.*

diversity: (having) distinctive multiplicity. Contrast *unicity.*

divine: spiritual excellence.

dobe (pronounced "due-bee"): doing & beĩng, the 2 modes of living.

doctrine: a philosophical conviction.

dogma: a codified belief system (creed).

doing: behaving; performing (a task). Compare *beĩng.*

double entendre: a word or expression with 2 possible interpretations.

dream: mentally generated perception while asleep. Compare *imagination.* See *daydreaming.*

dualism: the belief that existence has 2 realms: the physical and the mental (or spiritual). Contrast *monism.*

dullard: a stupid person.

dynamic (noun): a field process.

E

economy: the art of managing; frugality in use; selective exposition to facilitate understanding; organization of operation; a societal system of production, distribution, and consumption for bodily survival.

economics: conceptualization of a societal economy.

effect (verb): to cause to happen.

efficacy: the ability to produce desired results; effectiveness.

ego: sense of self, including self-esteem.

electromagnetic radiation: immaterial field travel in the electromagnetic spectrum.

electromagnetism: the interactions of electricity and magnetism, which appear distinct but are intimately related. Electromagnetism is one of the fundamental physics forces. See *gravity*.

electron: the fundamental particulate field of electromagnetism. As part of an atom, electrons orbit the atomic nucleus (core of nucleons).

emanate: to emit, issue, come or send forth.

embryo: an animal at an early stage of development.

embryogenesis: formation and development of an embryo.

emergence (philosophy): the synergistic construal that complexity arises from a multiplicity of simple interactions, as when a complex entity assumes properties or behaviors that its constituents do not have on their own, emerging only when they interact in a larger system.

emergence (physics): actuality becoming phenomenal on a Plank-time moment-by-moment basis.

emotion: a feeling developed by nattermind into a sustained mental state that disrupts awareness.

emotive: emotional or relating to emotion.

empathy: emotive mind perception.

empower: to enable.

endothermic: an energetic heat-absorbing chemical reaction. Contrast *exothermic*.

energy (physics): the idea of an immaterial force acting upon or producing matter. Energy may also be mentally felt as a vibrational sensation. Though energy is the foundational construct of existence, energy itself does *not* exist. As matter itself is made of energy, this tautology tidily proves *energyism*.

energyism: the monistic paradigm that Nature is a figment of the mind. Energyism differentiates between actuality and reality. Contrast *matterism*. See *dualism*.

enlightenment: the 1st level of higher consciousness, whereupon contented quietude reigns and bliss is a common occurrence. Contrast *iğnorance* (the lowest level of consciousness). See *cöherence consciousness, realization*.

enliven: to cause to be lively or joyous.

entanglement: distinct phenomena sharing an environment.

entertainment: fun.

entity: a perceived form or abstraction. *Entity* is a general term for anything which is construed as existing. The mind treats abstraction as existing in having a conceptual form.

environ: surroundings.

environment: a designated spatial region or conceptual realm.

equity: fairness.

eukaryote: an organism with cell structures (organelles) separated by membranes. Multicellular life is eukaryotic. Compare *prokaryote*.

event: a perceived process with an outcome.

evidence: an experience that gives confidence in conceptualization accuracy.

evolution (generally): the idea of vectored change through time.

evolution (biology): adaptation in a living being.

excitement: something which emotively agitates. Compare *amusement*.

existence: what appears in the mind as perceptual. See *actuality.*

exists: perceptible.

exothermic: an energetic heat-releasing chemical reaction. Contrast *endothermic.*

experience (noun): a sensory encounter or event.

experience (verb): to sense.

expression (philosophy): the act, process, or instance of manifesting.

extent: spread; magnitude; the space, range, distance, or expanse over which something has influence.

externality (economics): an unintended byproduct of an ecological activity, typically making an economic product. Waste and pollution are exemplary externalities.

F

fact: an experienced event.

faith: belief in absence of evidence.

fantasy: experience based upon desire, involving imagination.

fascicle: a localized bundle or compact cluster.

fate: the idea of cause-and-effect writ large.

feel (psychology): inner appraisal of contentment.

feeling: a perceptual reaction not (yet) developed into an emotion; a sensation not directly associated with the senses.

fermion: a quantized field which appears as matter. Compare *boson.*

fictional: nonexistent. Compare *truth.*

field: (typically localized) data with influence potential. A field may influence other fields and be influenced by other fields. Compare *gyre.*

fission (physics): breaking atoms apart, thereby releasing tremendous energy. Contrast *fusion.*

forage: search for food.

force (physics): the old-school, pushier physics term for *interaction*. Sometimes force fits, as some fields are organizers. See *hydration force*.

force (sociology): imposing one's will on others. Contrast *power* (sociology).

form: matter; a substance.

fortify: to strengthen or invigorate.

fortitude: strength, patience, and endurance in the face of adversity or temptation.

fractal: a self-similar pattern.

freedom (psychology, economics): contentment and emotional detachment; the mental state of release from materialism. Contrast *materialism*.

friction (physics): resistance to relative motion; inertia by proximity to other fields.

frustration: emotively feeling a hinderance to satisfaction.

fun: something that arouses mirth.

fusion (physics): the process of multiple atomic nuclei fusing. Contrast *fission*.

G

Gnosticism: various ancient religions whose adherents forsook the material world for spiritualism. Christianity is conceptually Gnostic. Christianity's corruption by the Catholic Church erased the Gnostic gospel, substituting faith and obedience as the main lessons.

God: the idea of a metaphysical, immortal supreme being who is omniscient and typically omnipotent.

gravity (physics): the idea that matter has an attractive force related to its mass; an attractive unipolar derivative cöforce of mass.

greatness (spirituality): feeling bliss in beĩng.

gullible: inclined to believe; easily tricked.

guru: a spiritual teacher in realization. Compare *sage*.

gyre: a rotating vortex; a conceptual framework treating a physical system as a dynamic interacting with its environment. Compare *field*.

H

habitat: the environment in which a life form lives; living conditions.

hallmark: a distinguishing characteristic or feature.

happiness (noun), *happy* (adjective): a moment or mood of feeling good. Compare *joy, bliss.*

harmony: accord with an environment; a pleasing or orderly relation among components; in music, simultaneity of notes causing a pleasing sonic blending.

heat: electromagnetic energy felt as warmth, as it is in the infrared region, which is just slightly less energetic than visible red light.

heat capacity (aka *thermal capacity*): the amount of heat needed to raise the temperature of a substance.

Higgs mechanism: the Planck-time process whereby gauge bosons acquire their attributes via spontaneous symmetry breaking. The Higgs mechanism exemplifies the basic mechanism by which Nature is composed: universal fields localizing, with local fields quantizing into perceptible particulates: objects coherently formed from the quantum level up.

homeodynamic: (the idea of) a self-sustaining system.

hominin: the hypothesized clade of ape that descended into humans.

hydration: a chemical compound that contains water.

hydration force: the field influence of water.

hypothesis: a guess. Compare *theory.*

I

idea: a concept imbued by imagination.

identity: sense of self.

ignorance: not knowing.

iğnorance: the level of consciousness where nattermind runs roughshod in mentation, causing suffering for the resident consciousness. The breve symbol ˘ (over ğ) signifies nattermind lording over the mind of those in iğnorance. Compare *enlightenment.*

imagination: the faculty for forming counterfactual mental images and perceptions. Compare *dream*.

immaterial: not having physical form; incorporeal; apparitional.

immortal: extending beyond a single lifetime.

immune system: a biological system that wards against disease, especially infection. For macrobes, an immune system acts as a microbiome management system.

imperceptible: not capable of being perceived.

impress: to make a strong impact on, especially positively.

inanimate: not endowed with life. Contrast *animate*.

incarnate: to make a mental appearance in Nature.

indelible: incapable of being removed.

induction (logic): the method of inferring a generalized conclusion from particulars. Contrast *deduction*.

inexplicable: unable to be explained by physical mechanics.

infection: an illness in an organism caused by invasion of another organism.

infer: to guess based upon observation.

infinite: beyond measure.

influence: an entity altering construal of another entity.

information: data given meaning; a perception considered of some value. Compare *data*.

ingrained (psychology): firmly impressed or fixed in the mind.

innate: inborn; knowledge or a means of acquiring knowledge within a being without learning. The ability to learn is innate.

insight: a fresh comprehension.

instantiate: to represent (an abstraction or principle) by an actual instance.

instinct: precocious knowledge.

integral: being an essential part of a whole; essential.

intelligence: behavior appropriate to a situation in that it may bring advantage.

intent: pure mental focus; intention.

intention: a purpose; a goal; a plan.

interaction (physics): a construal of functional relation between perceived entities. See *force*.

interplay: interaction and/or reciprocal influence.

intrinsic: inherent; being essential to or of the nature of something.

introspection: insight about mentation; construal of menition. Menition is mentation exclusive of perception.

intuition: instant comprehension without conceptualization. Insight is an intuitive experience.

inyō (Japanese: 陰陽; aka *yinyang* (Chinese)): complementary opposites. An inyō is a paradoxical dynamic.

irrefutable: impossible to refute or disprove; indisputable. See *truth*.

irony: a contrast that is unexpectable.

is (noun): the construal of an experience.

is (verb; an inflection of *be*): the process of experience.

J

jeopardy: felt risk of loss.

joy: exalted happiness. Compare *bliss*.

K

kinetic energy (physics): the energy of an active field, measured by motion. Compare *potential energy*.

knowledge: a grasp of concepts that may have utility; the idea that something in Nature works a certain way.

knowledgebase: the store of memory which is considered as knowledge.

L

learn: to gain knowledge via perception.

levity: a feeling of joy.

life: the idea of living as beings of form.

Life: the totality of living entities or applying to all life forms.

light: electromagnetic radiation which can be sensed by human eyes.

living: the process of being alive.

localize: the process of multiplying or diversifying a universal field. Compare *quantize*.

logic (academics): the study of reason.

logic (dynamics): the process of chaining concepts together – from premise (assumption) to conclusion (inference) – in a way that the linkages become agreeable. "Logic is invincible. To combat logic it is necessary to use logic." ~ French mathematician Pierre Boutroux.

logical: via conceptual reasoning.

love: adoration of a concept.

lucidity: clarity in awareness and undisturbed mentation.

M

macrobe: non-microbial life; any organism not requiring a microscope to be seen. Contrast *microbe*.

macromolecule: a large molecule.

madness: a mind severely corrupted and weakened by nattermind and malign presences.

magic: a mysterious process.

manifest (verb): strictly, to be sensed or perceived; broadly, to be involved with perception.

manifestation: a perceptible experience.

mantra: a vibrational meditation device to pacify monkey-mind into silence.

mass (physics): an extensive property of matter that measures its inertia and gravitational pull. At the ambient scale, mass and weight are much the same. But this simple relation is stressed in modern physics.

mass extinction event: in a geophysical period, abrupt, widespread demise of many species living on Earth.

materialism (psychology, economics): a worldview valuing material consumption and possessions. Contrast *freedom*.

materiality: belief in physicality as real.

matter: a field or proximate bundle of fields that a mind construes as physical. See *energy*.

matterism: the monistic belief that reality is made of matter. Matterism supposes that the mind is a figment of something substantial. Contrast *energyism*. See *dualism*.

meaning: mental significance assigned to a perception, idea, or emotion. Meaning is typically related to attraction or avoidance.

mechanical: pertaining to matter.

mechanics: (a) science pertaining to mechanisms.

mechanism: a definable process or function. See *parameter*.

meditation: pure awareness; a practice for silencing natter-mind and attaining transcendence.

menition: subconscious mentation exclusive of perception.

mental health: a level of consciousness above iğnorance.

mental illness: iğnorance.

mentalize (aka *mind perception*): inferring the mental state of someone else; imagining what someone else is thinking and/or feeling.

mentation: mental activity. See *perception*, *menition*.

metaphysical: incorporeal; transcendental to physicality.

method of exhaustion: a mathematical iterative method of finding the area of a shape by inscribing inside it a sequence of polygons whose area converges to the area of the containing shape. The method of exhaustion requires proof by contradiction, known as *reductio ad absurdum*. The contradiction involves repeatedly proving that the area of a proposed polygon is false (not equal) to the shape until falsification fails, thus leaving a true result. Proof by contradiction is freely used in mathematics, though the rigor of its logic is dubious, as proof is indirect.

methodical: done in an orderly and systematic way; marked by systematic, orderly, or unimaginative behavior.

microbe: a microorganism, too tiny to be seen without a microscope; often a single-celled prokaryote. Microbes include archaea, bacteria, and fungi. Contrast *macrobe*.

microbiome: the families of microbes that colonize plants and animals.

mind: the noun form of mentation. Minds don't exist.

mind perception (aka *mentalization*): see *mentalize.*

mind-body problem: the unsolvable inquiry into the functional interface between the intangible mind and the seemingly physical body.

miracle: inexplicable except via the doctrine of energyism.

mirage: an illusion of seeing something that is not there.

model (mathematics): a numeric simulation of a system.

molecule (physical chemistry): a configuration of atoms under sway of their electrons.

monkey-mind (aka *nattermind*): the involuntary, independent agent in the mind that instills and maintains iğnorance. Contrast *willmind.* See *coremind.*

monism: the doctrine that there is a singular reality, either matterism or energyism. Contrast *dualism.* See *energyism, matterism.*

monopole (*monopolar*): having a single pole. Compare *dipole.*

monotheism: belief in a singular god. Compare *polytheism.*

mood: an emotive frame of mind.

moral universe: the belief that morality exists in Nature.

morality: principles about right and wrong regarding relations with other life.

motion: a memory of something being in a certain location at one time and then being somewhere else.

mundane: ordinary because of abundance; commonplace.

music: sound perceived as patterned via repetitive elements.

mutualism: a regular interaction between 2 organisms that provides mutual benefits.

mystical: having hidden mechanism or meaning.

N

naïve: marked by simple-mindedness; unsuspecting; credulous.

naïve realism (philosophy): the belief that actuality as perceived is reality; that the senses provide direct comprehension of reality, and that reality is objective, subject to subjective interpretation. Both dualism and matterism are naïve realisms.

nattermind (aka *monkey-mind*): the involuntary, independent agent in the mind that instills and maintains iğnorance. Contrast *willmind.* See *coremind.*

natural philosophy: the study of Nature from a holistic perspective; the common method of comprehending Nature until the 17th century, before the scientific method took over. See *science.*

Nature: the expression of existence as construed by minds.

need: something necessary to stay alive. Compare *desire.*

negativity: an emanation of sour feeling. Contrast *positivity.*

nescience: lack of knowledge; ignorance.

nominal: in name only; for appearance's sake.

noumenon: outside of existence. A noumenon is beyond perception, as contrasted to *phenomena.*

now: the present moment; all that temporally exists.

nucleon: a subatomic particle in an atomic nucleus. Each atomic nucleus has 1 or more nucleons. Protons and neutrons are the 2 known nucleons.

nurture: take care of in a way that encourages development or growth of something with vitality.

O

object: something which appears to have a material (physical) form.

objectification: the mental creation of objects.

objectivity: the idea that Nature is independent of consciousness. Compare *showtivity.*

odorant: an odorous substance.

olfaction (aka *oflactics*): the act or sense of smell.

ordain: to grant as the nature of existence.

order (noun): discernible organization or arrangement.

order (verb): to command; to put in order (organize).

ordering (noun): a mode of arrangement, regulation, or management.

organism: an animated organic structure; a living life form, including single cells. Viruses are not typically referred to as organisms, as they lack the ability to reproduce.

organize: to set in order; arrange in a perceptible pattern.

oscillate: to go steadily and repeatedly back and forth. See *vibration*.

ouroboros (*uroboros*): traditionally, an ancient symbol of a dragon or serpent eating its own tail, symbolizing the eternal cycle of incarnation. In Canticle, a paradox resolving in a conceptual amalgamation.

out-of-body experience: reference frame sensation from outside the body. See *astral projection*.

outcome: a construed conclusion to an event.

oxidant (aka *oxidizer, oxidizing agent, electron recipient, electron acceptor*): a substance in a redox chemical reaction that gains an electron from a reductant (reducer, reducing agent, or electron donor). An oxidant is any substance that oxidizes another substance.

oxidation: an increase in oxidation state via loss of electrons. Contrast *reduction*.

oxidation state (aka *oxidation number*): the hypothetical degree of oxidation (loss of electrons) of an atom in a chemical compound. Conceptually, oxidation state may be zero, positive, or negative. Oxidation state is a useful predictor of electrical charge.

P

paracosm: an imagined world.

paradigm: a way of understanding.

paradox: a seeming conceptual contradiction which may nonetheless be reconciled by construing a complement between the concepts.

parameter: an expression which determines the form or behavior of something.

paranormal: immaterially phenomenal; beyond verification from matter-based evidence.

particle: a portion or division of a whole.

particle (quantum physics): the idea that the quanta that comprise the building blocks of Nature are itty bits. See *quantum*, *field*.

particulate: existing in the form of distinct particles.

passage: a statement in a written (or musical) work.

pattern: (perception of) a distinctive design.

perceptibility: capable of being perceived.

perceptible: within potential perception.

perception: the process of making sense of sensation, including determining significance and meaning.

perceptivity: psychological strength; (degree of) keenness in perception.

perceptual: related to sensory perception.

perspective (psychology): a mental reference frame.

perspicuity: the quality of being clear to understanding; mental clarity.

pervasive: widespread; present everywhere.

phantom: an illusory appearance or figure.

phenomenon (plural: *phenomena*): subject to experience; perceptible. See *actuality*. Contrast *noumenon*.

philosophy: a set of consistent definitions for a system. The term *philosophy* derives from ancient Greek for "love of wisdom."

photon: a quantum of light; the theoretical cöforce of electromagnetism.

physical: sensation of objects and perception of matter.

physical system: a construed collection of objects. A being (life form) is a physical system by being an amalgamation of cells or other components.

physics: the study of matter and its movements. See *classical physics*, *modern physics*.

pity: compassion.

Planck constant (aka *Planck's constant, Plank's action quantum*): a physical constant reflecting the size of energy quanta in quantum field theory. Planck's constant states the proportionality between the momentum and quantum wavelength of every subatomic particle. The relation between the energy and frequency of quanta is the *Planck relation.*

Planck length: the minimal theoretical limit to spatial distance, derived from Newton's gravitational constant, the speed of light in a vacuum (*c*), and Planck's constant. Planck length is 1.616199×10^{-35} meters.

Planck mass: the theoretical amount of mass in a sphere with a radius of Planck length, with a density of 1093 g/cm^3.

Planck time: the theoretical limit of temporal measurement; the time required for light in a vacuum to travel a single Planck length: 5.391×10^{-44} seconds.

Planck unit: a system of natural units used in physics, particularly *Planck length* and *Planck time.*

plasmon: a quantum of vibration (oscillation).

pleasure: a feeling of happiness or satisfaction.

plutocracy: conservative political rule by the wealthy. To preserve the privileges of the rich, conservative rule aims to maintain the status quo.

polar (philosophy): oppositional.

polarity: the idea of opposites, which may be expressed in a spectrum or by tension between poles. Electromagnetism functions via polarity in both senses of the word.

political economy: an economic system supported by the authorities (political regime). Plutocratic capitalism is the dominant political economy throughout the world.

polymath: someone of encyclopedic learning.

polytheism: belief in multiple gods. Compare *monotheism.*

positivity: an emanation of good feeling. Contrast *negativity.*

potent: producing a strong effect; influential.

potential energy (physics): the inherent or potential power in a field. Compare *kinetic energy.*

power (physics): relative ability to affect or effect.

power (psychology): the efficacy of willmind expression. Compare *strength*. See *power* (sociology).

power (sociology): influence on others. Contrast *force*.

predictable: foreknowable; expectable; capable of being known in advance.

pregnant (philosophy, not biology): full or fraught.

presence (biology): an incorporeal being, typically referring to a mental interloper, though a presence is not necessarily a mind inhabitant. A presence is sometimes called a *spirit*, which is a more generic term.

primal: original; fundamental.

principle: a conceptualized claim of fundamental truth.

process: a sequence of motions conceived as related. See *event*.

product (philosophy): an outcome. Compare *artifact*.

prokaryote: an organism that lacks a cell nucleus or other membrane-bound organelles, though prokaryotes are structurally compartmentalized. While prokaryotes are strictly single-celled, most form stable, aggregate communities, such as a biofilm. Compare *eukaryote*.

proper: appropriate.

property (physics): a characteristic, attribute.

protein: a living macromolecule from which cells are made and the workforce which keeps cells alive; alternately used to describe foodstuffs heavy in molecular protein.

proximal: near, close. Compare *distal*.

psychic: relating to psychological paranormal.

psychology: the study of the mind.

psychospace: the mental world of space and time.

purpose: a motive or intention; resolve; a reason for existence.

Q

quality: distinguishing characteristic or feature.

quantize: (seemingly) fragment into distinct bits (typically particulates) from a unitary field via field localization. Compare *localize*.

quantum (plural: *quanta*): a localized energy field that appears particulate via quantization.

quantum field theory: theory about how existence is fabricated from the quantum scale up via fields.

quantum mechanics: a mechanistic view of Nature coming into existence via quanta assemblage. See *physics, quantum, quantum field theory*.

quietude: inner silence. See *enlightenment*.

R

radiation: (electromagnetic) field travel.

rage: intense anger; fury.

random: unpredictable; stochastic; purposeless.

react: respond (to an influence or stimulus). Contrast *act*.

real: true. See *reality*.

reality: that which necessarily is, regardless (and outside) of existence.

realization (aka *self-realization, unity consciousness*): the highest level of consciousness, where an individual being communes with Cönsciousness.

realm: domain, region.

reasoning: making sense of experience or concepts by applying logic.

recognize: attribute to a certain category by memory recall and selective pruning of outlying characteristics.

redox (*reduction–oxidation*): a chemical reaction in which the oxidation state of a substance changes via electron transfer. Reduction is an electron gain whereas oxidation is an electron loss.

reductant (aka *reducer, reducing agent, electron donor*): a substance in a redox chemical reaction that loses an electron to an oxidant.

reduction (chemistry): a decrease in oxidation state via a gain of electrons.

reference frame: mental perspective.

refraction (physics): changing the direction or angle of travel of a radiating field.

refractive index: a geometric indication of the angle an electromagnetic radiative field bends when refracting between media. A *medium* is a substance that a field is traveling through.

reify: to regard an abstraction as existing.

reincarnation: the incarnation of a soul in someone who remembers a previous incarnation. See *transmigration*.

relative: having meaning only by contrast. Contrast *absolute*.

religion: a shared paradigmatic belief system about the nature of reality. Religions are frequently faith-based, typically dogmatic, and usually involve supernatural agents (gods). Compare *natural philosophy*. Contrast *science*.

resolve: firm intention.

retina: the light-sensitive layer of tissue lining the inner surface of the eye.

rich: having attained full potential of dobe, positivity, or productivity.

risk: imagined loss potential.

riveting: having the power to compel attention; fascinating.

rulebase: a set of rules applicable to a system.

S

sage: a person in realization, the highest level of consciousness. Compare *guru*.

saltation (biology): a seemingly abrupt evolutionary change from one generation of organism to the next.

sans: without.

satisfaction: gratification; enjoyment of an outcome.

science: knowledge; also used for the study of Nature. As evidence is material, the term *science* is commonly used under the assumption of dualism or matterism: a naïve religious belief which betrays the spirit of science. Compare *natural philosophy*.

self: construal of the identity of spirit. In iĝnorance, self is identified as an aware mind-body amid an external world (typically conceived as the naïve realism of dualism). At higher levels of consciousness, self is identified as a consciousness witnessing a showtivity. See *spirit*, *showtivity*.

self-esteem: pride in oneself.

self-realization: See *realization*.

selfsame: identical.

sensation: uptake from the senses. See *perception*.

senses: bodily organs, tissues, or cells which detect physical existence or motion.

sensible: capable of being sensed.

shell (physics, chemistry): an electron orbital layer.

showtivity: subjective witnessing of Nature (the "show") which may be similarly witnessed by others; the principle that the seeming objectivity of Nature is actually a shared subjectivity. Compare *objectivity*.

simultaneous: at the same time.

skepticism: questioning of a conceptual construct; disbelief.

skill: the ability to overcome a challenge.

socialism: an economy aimed at equity. To avert externalities, socialist theorists typically favor a command economy. Contrast *capitalism*.

society: a group of people who live in the same geographical region and are linked by authority and customs. See *tribe*.

soul: an immortal spirit which incarnates as a consciousness.

space: the set of dimensionality which yields the appearance of Nature. See *time*.

spacetime: the idea that space and time are dimensions of existence.

species (science): a grouping via induction.

spectrum: a relative range of quality or conceptualization about some system.

spirit (parapsychology): the essence of consciousness imbued with a life force; an incarnate soul. Not all spirits take bodily form. See *presence*.

spirit (psychology): vitality in living which is reflective of willmind. Compare *ikigai*.

spiritbiome: the paranormal spirits that colonize a mind.

spiritualism: a synonym for *animism*.

spirituality: (related to) consciousness level, especially leveling up awareness; freeing oneself from the clutches of monkey-mind.

squaring the circle: a metaphor for trying to do the impossible. The idiom refers to a geometry problem that originated in ancient Greece: defining a square with the area of a circle.

stage: a level, step, or period in a process, activity, or development.

standing wave (aka *stationary wave*): a coalition of intersecting fields that sustains itself as an emergent field.

stratagem: a plan to trick or deceive; a deceitful trick; a ruse.

strength (psychology): perceptivity; the operational state of consciousness level; relative clarity. Compare *power* (psychology).

stress: mental suffering.

strive: to work hard; to struggle.

strong force: the cöforce that binds together the nucleons of an atom's nucleus.

stupidity: inapt behavior from dim awareness. "Stupid is as stupid does," was an idiom taught by American mothers to their children in the early 19th century as a retort to bullying.

subconscious: beneath conscious awareness. Compare *conscious*.

subjective: the perspective of an individual mind.

subsistence: in existence.

substance: matter; something tangible.

suffer: to experience pain or distress.

superconductivity: zero electrical resistance, resulting from electrons overcoming their mutual repulsion and pairing up, creating a coherent, frictionless flow.

survey (noun): comprehensive overlook, including extent and quality.

suspended animation: an inert life form that retains vitality.

suss (chiefly British): figure out; (seek to) comprehend.

sway: influence.

symbiont: an organism that lives symbiotically with a host.

symbiosis: 2 dissimilar organisms in continual interaction, often in a mutually beneficial association (mutualism).

symbol: an abstraction that signifies something; a datum which may acquire meaning; a conceptual element.

system: a conceptual construct of an amalgamation.

systematic: nonrandom behavior following certain principles.

T

takeaway: a key point(s) or conclusion from a collage of information.

tangible: capable of being perceived by touch.

task: work that has a set goal.

temporal: relating to time.

tenet: a doctrine considered true.

tension: electromagnetic potential.

theme: a topic or subject.

theory: an explanation based upon evidence. Compare *hypothesis*, *truth*.

theory of mind: the recognition that everyone has their own unique mind. See *mind perception*.

thermodynamics: the physics of heat related to energy.

thought: conscious attention to symbols and concepts.

thrall: bondage or slavery.

time: the idea of change. The existence of time is purely in the mind, as memory (the past) and imagination (the future). See *space*.

topology: the study of conceptual regions and fields.

trait (biology): a purposeful, often functional, attribute.

transceive: share (receive and transmit) data; communicate.

transcendence (philosophy): direct apperception beyond the confines of concepts.

transcendence (biology): the state of consciousness where the mind is quiet while the body is receptive to sensory experience.

transcendental: surpassing the limits of actuality.

transform: to change the nature, character, function, or condition of.

transmigration: passage of consciousness in a soul from one incarnation to the next. Compare *reincarnation*.

transpire: (to) occur; happen.

tribe: a social group with a shared culture.

true: real, genuine, authentic. See *truth*.

trust: confidence in the reliability of something or someone.

truth: a construal that cannot be contradicted by evidence or inference from evidence. Compare *theory*.

U

underlie: to be the foundation or basis of.

understanding: conceptualization that satisfies curiosity.

unicity: oneness. Contrast *diversity*.

unitary: of a unity.

utility: of some use.

V

valid: applicable, well grounded, reliable.

value (psychology): worthy of esteem, often by recognition of having productive use.

vanquish: subdue; subjugate; overcome; subordinate.

veracity: conformity with truth.

vibrate: to quickly oscillate. Vibration is the coherent interactivity of every field.

virus: a molecular parasite.

vital: necessary to staying alive; more generally, essential.

vitalism: the principle that life has a distinct vital energy. Compare *animism*.

vitiate: (to) mar, taint, invalidate.

volition: an act of will.

votive: done with dedication.

W

wave: (a) radiating field; the idea of a coherent, energetic, localized traveling field. Physicists may use *wave* rather than *field* to emphasize vector, magnitude, and frequency rather than range of influence. See *field*.

wile (noun): a trick/artifice/stratagem of deceit.

will: (strength of) intention.

willmind: volitional mentation. Contrast *nattermind*. See *coremind*.

wisdom: apperception of truth.

witness: intentionally experience.

work (physics): energy in transit; the outcome of an energetic force applied to matter.

work (psychology): effort toward an outcome.

world (the): Nature perceived.

Y

yield (verb): to give forth or produce.

About the Author

Ishi Nobu is a guru and polymath scholar. His scientific revelations stem from holistic comprehension, drawn from extensive study. His spiritual teachings follow the tradition of sages since antiquity, who have taught a selfsame lesson about how to make the most of one's life.

In his oeuvre, Nobu creates a consilience of the sciences with his accounts on the dynamics of Nature and humanity.

In *A Canticle of Reality*, Nobu employs a poetic technique used in archaic spiritual texts, guiding the reader to the intricate interconnectedness of reality, existence, knowledge, and mental health.

Nobu's work is available through ***IshiNobu.com***.

Index

Made in the USA
Middletown, DE
21 August 2023

37105827R00076